PERSPECTIVES OF OIL AND GAS: THE ROAD TO INTERDEPENDENCE

Perspectives of Oil and Gas: The Road to Interdependence

by

Marcello Colitti
with **Claudio Simeoni**
Fondazione Eni Enrico Mattei,
Milan, Italy

KLUWER ACADEMIC PUBLISHERS
DORDRECHT / BOSTON / LONDON

A C.I.P. Catalogue record for this book is available from the Library of Congress.

ISBN 0-7923-4086-8

Published by Kluwer Academic Publishers,
PO Box 17, 3300 AA Dordrecht, The Netherlands.

Kluwer Academic Publishers BV incorporates
the publishing programmes of
D. Reidel, Martinus Nijhoff, Dr W. Junk and MTP Press.

Sold and distributed in the United States and Canada
by Kluwer Academic Publishers,
PO Box 358, Accord Station, Hingham, MA 02018-0358, USA

In all other countries, sold and distributed
by Kluwer Academic Publishers Group,
Distribution Center, PO Box 322, 3300 AH Dordrecht, The Netherlands

Printed on acid-free paper.

Printed and bound by Hartnolls Ltd., Bodmin, Cornwall.

CONTENTS

FOREWORD

Marcello Colitti and Claudio Simeoni present a thoughtful and incisive analysis of the development and dynamics of the global oil business in their new book, *Perspective of Oil and Gas: the Road to Interdependence*. The book captures the themes of the business and the interests and needs of both producers and consumers – and the companies that provide the essential services to both.

I have known Marcello Colitti for more than two decades, and have great respect for his knowledge, wisdom and judgement – a respect that is shared throughout the industry. He brings the unique perspective of participant and expert through the many decades. His unique experience in ENI, as it has achieved and sustained its great role, adds to his authority.

Marcello Colitti is a man of rigorous analysis. But he also has, as members of the industry throughout the world know, more than a touch of the artist about him – a special 'sensibility for oil'.

We are fortunate that he and his knowledgeable and expert colleague Claudio Simeoni can now apply their knowledge – and their distinctive perspective – in this contribution to the global thinking on the oil industry as it is today – and will become tomorrow.

Daniel Yergin
President, Cambridge Energy Research Associates
Author, *The Prize: the Epic Quest for Oil, Money, and Power* (Pulitzer Prize Winner)

M. Colitti and C. Simeoni, Perspectives of Oil and Gas: The Road to Interdependence, vii.

PART I

THE RESOURCES

CHAPTER 1

THE SIZE OF PRESENT AND FUTURE HYDROCARBON RESERVES

Introduction

On 3 May 1866, while discussing the Budget, the Chancellor of the Exchequer, Gladstone[1], asked the reason for the "commanding position" of the British economy, and announced that "the chief cause is our mineral treasures . . . . the possession of vast stores of coal . . . . that we can raise . . . . to the surface at a lower price than any country in the world". Coal was the "raw material of the metal manufacture, especially the gigantic business of the iron trade".

Gladstone also asked a more precise question: "Assuming for the moment that we shall not be able to continue for many generations to produce coal at prices cheaper than those of all other countries, what will happen?".

According to one of their most famous past statesmen the industrial countries already had, in the nineteenth century, a clear idea of the relation between their "commanding position" and their supply of cheap energy; and started worrying very early about it. Of course, that position lasted, for Great Britain, for a shorter time than the supply of British coal, which was still abundant in the mid-twentieth century.

Today, the industrial countries are not the only ones who worry about the lifespan of their energy supplies. The oil-exporting countries, many of which are on the way to industrialisation, worry even more. Their strategy of depleting their reserves, and therefore their supply strategy, must be defined on the basis of the duration of such reserves.

How long are hydrocarbon (that is, oil and gas) reserves going to last? What is the best depletion policy? And what is the best level of prices?

This book will try to provide some answer to these questions, and to some more. In Part Two it will also describe the present structure of the oil industry, its main actors and the complex development which led to it. The final chapter will

---

[1] *Hansard* (vol. 285) May 3 1866, col. 365–403: Chancellor of the Exchequer (Rt. Hon. W.E. Gladstone) – The Budget – Financial Statement. See col. 399f. Obtained by courtesy of Bodleian Library, University of Oxford.

Jose Harris, in *Private Lives, Public Spirit. A Social History of Britain 1870–1914*, Oxford University Press 1993, p. 256, says that Gladstone "had struck chill into the hearts of the House of Commons by a speech which suggested that Britain's economic ascendancy would last just as long as her supplies of coal – provisionally estimated to be sufficient for another hundred years".

M. Colitti and C. Simeoni, Perspectives of Oil and Gas: The Road to Interdependence, 3–27.

attempt to define the role of oil policy in overall economic development policy through an approach which will make the most of the interdependence between oil exporters and oil importers.

First, however, we will answer the question which worried the English Chancellor in 1866: how long will the supply of energy last? What will be its effect on our economic life?

Oil and gas reserves, proven and to be found

Oil underground is "contained" in a field, a solid whose volume, shape, inside pressure and geological nature are estimated through correlations among the points reached by exploratory wells. The volume of oil it may contain ("oil in place") is estimated through indirect indicators, basically its volume and pressure, and the porosity of the sands or rocks which trap the oil. In the same way the expected percentage of recovery of the oil in place (usually rather less than 50%) and the production level are estimated by calculating how the pressure of the field will behave according to the number of wells drilled and the production per well. These are calculations on uncertain data, and surprises, both nice and nasty, are frequent. Moreover, production does not depend on technical factors alone: the demand for oil and the oil producer's financial needs also have to be taken into account.

Of course, initial estimates are reviewed systematically. Reconsidering the original evaluation, usually with improved instruments and methodologies, the oil producer may find good reasons for drilling more wells; if he is lucky he may find that there is more "oil in place" than previously estimated.

Everything in this process is based on indirect knowledge gathered about a field over many years of work covering all the different phases of a complex technology: seismography, exploration, appraisal, production planning, etc.

A similar procedure can be applied in order to evaluate the total reserves that may exist in the sedimentary basins of a certain area, or in the world; that is, to estimate the amount of hydrocarbons which has not yet been found, but could be found in the future. In this case the indications given by the existing wells are supplemented by others gained through "regional geology", general parameters obtained from the study of hydrocarbon basins. This kind of study has been developed by many authors, and their results, although not certain, are in reality quite credible.

The estimates of how much oil is to be found in the world now make quite a long list (see Figure 1.1) including the estimates published by the authors of this book working with ENI–AGIP. The total volume of estimated world reserves has grown slowly from 1940 onwards. The more recent evaluations significantly converge towards a narrowing band of values. Several factors account for this phenomenon:

Figure 1.1 Estimates of world ultimate crude oil recovery

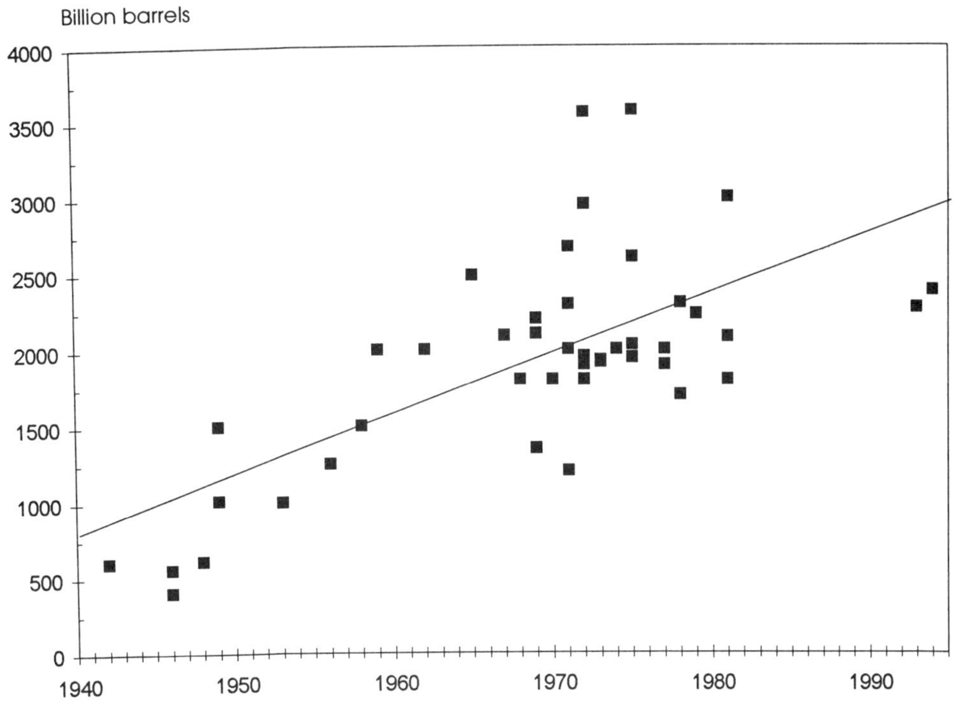

- progress has been made in the understanding of the geological and naphthogenic characteristics of the "provinces" already in production;
- more geological data have become available;
- the huge amount of technological progress in geophysical exploration has led to an extension of the prospecting for oil to the whole earth, albeit in a non-homogeneous way;
- virtually all the potentially productive areas in the world, including those that have not yet been investigated, or have been only partially investigated, have been classified in a number of clearly defined basins;
- methods of evaluating reserves have been improved.

The increasing importance of natural gas as an energy source does mean that gas reserves must be added to the evaluation of total world reserves.

Of course, many fields contain both crude oil and natural gas, mixed in very variable proportions; and some fields contain only gas. Although considered for long time a "dry well" a natural gas find is today quite valuable, as gas is not only the favourite fuel of the ecologists, but has also shown a great flexibility of uses. The following additional points should be noted regarding gas:

Estimates of world ultimate crude oil recovery

Date	Estimator	Organisation	Billion barrels
1942	Pratt, Weeks and Stebinger		600
1946	Duke		400
1946	Pogue		555
1948	Weeks		610
1949	Levorsen		1,500
1949	Weeks		1,010
1953	MacNaughton		1,000
1956	Hubbert		1,250
1958	Weeks		1,500
1959	Weeks		2,000
1962	L.G. Weeks	Consultant	2,000
1965	T.A. Hendricks	USGS	2,480
1967	W.P. Ryman	Esso (Exxon)	2,090
1968	–	Shell Oil Company	1,800
1969	M. King Hubbert	National Academy of Sciences	1,350–2,100
1969	L.G. Weeks	Consultant	2,200
1970	J.D. Moody	Mobil	1,800
1971	H.R. Warman	BP	1,200–2,000
1971	Weeks		2,290
1971	U.S. National	Petroleum Council	2,670
1972	J.D. Moody and H.H. Emmerick	Mobil	1,800–1,900
1972	Richard L, Jodry	Sun	1,952
1972	Linden		2,950
1972	H.R. Warman	BP	1,800
1972	Weeks		3,560
1973	Wim Vermeer	Shell Oil Company	1,930
1973	H. R. Warman	BP	1,915
1974	J.D. Moody and R.W. Esser	Mobil	2,000
1974	M. King Hubbert	USGS	2,000
1975	J.D. Moody and R.W. Esser	Mobil	2,030
1975	–	Exxon	1,945
1975	B. Grossling	USGS	2,600–6,500
1975	P. Odell	Erasmus University Rotterdam	3,575–4,233
1977	M. King Hubbert	Congressional Research Service	2,000
1977	–	World Energy Conference	1,900
1978	Richard Nehring	Rand Corporation	1,700–2,300
1979	A.A. Meyerhoff	Consultant	2,230
1981	Colitti	AGIP	2,082
1982	–	Exxon	1,800–3,000

Source: Timothy Greening, *Energy Economist*, September 1982, Issue 11.

The last two data points of Figure 1.1 – those in the years 1993 and 1994 – represent two estimates which are not included in this list: the one done in 1993 by C.D. Masters of the US Geological Survey and the one done in 1994 by the authors.

- A distinction must be made between associated and non-associated gas, since in the former case production together with oil is unavoidable.
- In some provinces containing exclusively gas, prospecting has been carried out to establish the potential – and in some instances the indicated potential is very high. Due to the remoteness of some of these provinces from the main centres of consumption, and the gaps that exist in the production–transport–international marketing chain, most of these provinces are not developed.

Estimate of reserves to be discovered

An assessment was made of the probability of finding new reserves by extending basin exploration area-wise and depth-wise. The size of the sedimentary basins onshore and offshore – that is, the aggregate total of areas where hydrocarbon deposits could exist – was measured for each region. Then the intensity of area-wise exploration performed in the basin was assessed. To establish this intensity – the "maturity" of a basin from the point of view of area-wise exploration – a criterion was adopted based on the ratio of the number of the new field wildcats (NFW) drilled and the extent of the potential basins. The figures obtained were then referred to a scale of coefficients of geographical exploration "maturity" in terms of the number of wells drilled per unit of sedimentary area. These coefficients represent the probability of finding new reserves by extending the search within the area. In other words, it is assumed that the reserves still to be discovered are inversely proportional to the exploration already done, due account being taken of the quantities found to date. The figures were modified for the Middle East, a basin of exceptional size with very large, scattered oil-pools.

The possibility of finding new oil reserves at greater depths was also considered, using the average depth of the NFW to establish the depth already probed by wells and comparing the results with a conventional scale which establishes "coefficients of vertical exploration maturity". As in the previous case, the vertical component of undiscovered reserves was computed in proportion to those already discovered, but in terms of the specific depth coefficient. Here too, however, it was necessary to adjust the purely statistical calculations on the basis of geological considerations concerning known basins. This is the case, for instance, in the US, where a multitude of shallow NFW hide a considerable number of deep ones.

To sum up, an evaluation was made of the extent and thickness of the sedimentary areas of the basins, and an estimate was made of their "maturity" from the point of view of exploration – that is, how much exploration has been carried out and the results achieved.

Table 1.1
OIL FOUND FROM THE BIRTH OF THE OIL INDUSTRY UP TO THE YEARS 1981, 1986 AND 1993 (in billion barrels)

Oil	*1981*	*1986*	*1993*
Proven reserves found	1,091	1,223	1,698
Reserves already produced	449	526	699
Residual proven reserves	642	697	999

Table 1.2
ESTIMATES MADE IN THE YEARS 1981, 1986 and 1993
OF WORLD OIL STILL TO BE FOUND UP TO THE YEAR 2030
(in billion barrels)

Oil	*1981*	*1986*	*1993*
To be discovered	445	412	387
Re-evaluations and extensions	202	180	128
Enhanced recovery	360	286	176
Total	1,007	878	691

The results in terms of reserves of oil which could be found in every area were published in 1981 in the Fall number of the *OPEC Review*, and later presented at the 12th World Energy Congress in 1983. Since that time the forecasts have been kept up to date.

Table 1.1 gives the estimates of reserves of oil found since the inception of the oil industry up to 1993, showing those already produced and those still available for production. Table 1.2, still related only to oil, estimates the reserves which could be discovered in the future up to the year 2030. It includes the reserves to be found, and also the re-evaluations and extensions of the existing fields, and what could be produced from the reserves by more advanced methods, which would increase the percentage of recovery, but apply only to oil. Table 1.3 puts all this together, and shows the total reserves discovered but still to be produced, the ones to be discovered, and the amounts coming from re-evaluations and enhanced recovery. Tables 1.4–1.6 repeat the procedure for natural gas. Finally, Tables 1.7 and 1.8 present the results for both oil and gas.

[2] A trillion is a figure of 1 with 15 zeros

Table 1.3
ESTIMATES OF WORLD CRUDE OIL RESERVES BY REGIONS FROM 1993 TO 2030
(in billion barrels)

Regions	*Resources*						
	Proven at end 1993		*To be discovered*	*Re-evaluations and extensions of new and old fields*	*Enhanced recovery in new and old fields*	*Ultimate reserves: plus additions up to year 2030*	
	(a)	*%*	*(b)*	*(c)*	*(d)*	*e = a+b+c+d*	*%*
Europe	18.8	1.9	15	6	4	43.8	2.6
Middle East	662.9	66.3	99	40	79	880.9	52.1
Africa	62.0	6.2	26	14	17	119.0	7.0
USA and Canada	28.8	2.9	58	19	20	125.8	7.4
Latin America	125.0	12.5	56	13	15	209.0	12.4
Far East	44.6	4.5	42	12	9	107.6	6.4
Russia and other CIS	57.0	5.7	91	24	32	204.0	12.1
World total	999.1	100.0	387	128	176	1,690.1	100.0

Note: Figures in the "re-evaluations and extensions" and "enhanced recovery" columns are estimated on the basis of the technologies that will be applied up to 2030, and an upper limit price of crude oil has been established at $21 per barrel.

Table 1.4
NATURAL GAS FOUND FROM THE BIRTH OF THE GAS INDUSTRY UP TO THE YEARS 1981, 1986 AND 1993
(in billion cubic metres)

Natural gas	*1981*	*1986*	*1993*
Proven reserves	109,050	134,804	195,493
Already produced	27,669	34,381	49,575
Residual proven reserves	81,381	100,423	145,918

Table 1.5
ESTIMATES MADE IN THE YEARS 1981, 1986 AND 1993 OF WORLD NATURAL GAS STILL TO BE FOUND UP TO THE YEAR 2030
(in billion cubic metres)

Natural gas	*1981*	*1986*	*1993*
To be discovered	85,280	94,980	110,200
Re-evaluations and extensions	47,080	46,140	45,300
Total	132,360	141,120	155,500

Table 1.6
ESTIMATES OF WORLD NATURAL GAS TO BE DISCOVERED BY REGIONS

Regions	*Billion cubic metres*	*Percentage*
Middle East	16,500	15.0
USA and Canada	13,500	12.2
Latin America	9,100	8.3
Asia and Oceania	7,600	6.9
Africa	8,900	8.1
Europe	6,300	5.7
Russia and other CIS	48,300	43.8
Total	110,200	100.0

It emerges that reserves still to be found amount to 691 billion barrels of oil and 155.5 trillion cubic metres of natural gas, broken down as follows:

- 387 billion barrels of oil and 110.2 trillion cubic metres of natural gas of new discoveries using present technologies;
- 128 billion barrels of oil and 45.3 trillion cubic metres of natural gas to come from re-evaluations and extensions of old fields;
- 176 billion barrels of oil to come from enhanced recovery.

Adding the known resources – produced and proven – total recoverable reserves could amount to about 2,389 billion barrels of oil and 351 trillion cubic metres of gas, of which respectively 699 billion barrels and 49.6 trillion cubic metres have already been produced.

The painstaking methodology used involved counting reserves field by field,

Table 1.7
ESTIMATED HYDROCARBON RESERVES IN THE WORLD IN THE YEAR 1993

	Natural Gas		*Oil*
	Trillion cubic metres	*Billion barrels of oil equivalent*	*(billion barrels)*
Reserves found up to 1993	195.5	1,290.3	1,698
less already produced	49.6	327.2	699
To be produced (*a*)	145.9	963.1	999
Additional reserves up to 2030 (*b*)	155.5	1,026.3	691
Total (*a* + *b*)	301.4	1,989.4	1,690
Total reserves found and to be found up to the year 2030	351.0	2,316.6	2,389

Table 1.8
RESERVES ALREADY PRODUCED AND ESTIMATES OF THE RESERVES TO BE FOUND UP TO THE YEAR 2030

	Natural Gas			*Oil*		
	Trillion cubic metres	*%*	*%*	*Billion barrels*	*%*	*%*
Total reserves found and to be found	351.0	100.0	100.0	2,389	100.0	100.0
Of which:						
– already found	195.5	55.7	–	1,698	71.1	–
– to be found	155.5	44.3	–	691	28.9	–
Of which:						
– already produced	49.6	–	14.1	699	–	29.3
– to be produced	301.4	–	85.9	1,690	–	70.7

and making detailed evaluations of the sedimentary areas of the world. However, it does not include the discoveries that might result using completely new technologies, or from unknown areas such as the polar icecaps. The results do not therefore represent the world's entire oil potential.

According to the figures, the quantity of oil which can be found between 1993 and the year 2030 is about the same as has been produced up to now; to this the present residual proven reserves must be added, which are 43% higher than the volume already produced. The whole picture is hardly one of scarcity: by the year 2030 1,690 billion barrels could be produced, more than twice the oil consumed from the origin of the oil industry up to the year 1993. The comparable figure for natural gas amounts to 301 trillion cubic metres, more than six times the gas consumed from the origin of the gas industry. The oil and gas still to be discovered up to the year 2030 may be broken down among the various areas of the world as reported in Tables 1.3 and 1.6.

Exploration needed to discover the potential oil reserves[3]

The petroleum provinces

Modern exploration activity – that is, seismic methods and drilling – has attained a high level of efficiency, at least for discovering giant fields. It can generally be assumed that, if the search for new fields is performed with efficiency, the "giants" would be brought to light within the first 7–10 years. On the other hand, there is little likelihood of a province becoming an important producing basin if the first 20 to 30 new field wildcats are dry, assuming that the well sites are chosen after careful geological and geophysical surveys.

About 600 petroleum-producing provinces have been identified in the world to date. Some are very large, such as the West Siberian Basin, and some are small. Moderate or intensive exploration has been carried out in about 450 of them; of these, 150 have been found dry or virtually so. Among the producing basins the Arab–Iranian one is clearly unique. It contains more than half of the total world reserves, while no other province has more than 10% of them.

Most of the oil consumed in the world today comes from the "20 major provinces" which contain 89% of the world's oil in terms of total recovery, while a further eight provinces contain around 5% and some 270 the remaining 6%. It takes between 6 and 15 years to establish the existence of resources in a province. An expansion in the total amount of oil resources will depend mainly on the discovery of new provinces classifiable as majors.

---

[3] The methodology of this chapter is fundamentally the same used in the article of *OPEC Review*, Fall 1981. Recent data have been calculated following the same procedure

Table 1.9
WORLD – EXCLUDING RUSSIA AND OTHER CIS – NEW FIELD WILDCATS DRILLED FROM 1961 TO 1992
By region
(number of wells)

Year	*Europe*	*Middle East*	*Africa*	*U.S.A./ Canada*	*Latin America*	*Far East China and Oceania*	*Total*
1961	297	26	268	7,879	224	40	8,734
1962	307	37	314	7,745	200	79	8,682
1963	263	28	365	7,572	209	106	8,543
1964	278	35	284	7,612	180	159	8,548
1965	237	44	281	7,102	217	149	8,030
1966	222	51	199	7,138	224	108	7,942
1967	179	53	207	6,271	210	91	7,011
1968	185	50	233	6,180	180	106	6,934
1969	238	49	225	6,946	188	167	7,813
1970	193	68	233	6,083	242	198	7,017
1971	206	63	238	5,476	255	201	6,439
1972	202	57	214	6,163	292	232	7,160
1973	239	74	171	6,454	225	251	7,414
1974	234	61	177	6,797	193	244	7,706
1975	305	84	196	7,192	220	232	8,229
1976	267	102	167	7,403	194	215	8,348
1977	240	82	211	7,810	217	171	8,731
1978	283	106	198	8,423	209	235	9,454
1979	216	80	164	8,335	231	275	9,301
1980	255	98	176	8,714	562	279	10,084
1981	316	118	234	9,685	462	349	11,164
1982	380	133	219	9,280	485	392	10,889
1983	359	105	190	7,502	392	394	8,942
1984	380	132	161	7,797	357	423	9,250
1985	417	121	218	7,040	331	447	8,574
1986	399	186	106	4,498	237	286	5,712
1987	282	129	167	4,591	247	345	5,761
1988	214	136	150	4,367	268	368	5,503
1989	345	77	167	3,701	239	313	4,842
1990	399	103	173	3,699	205	379	4,95
1991	369	154	168	3,200	212	405	4,508
1992	269	157	159	2,387	181	328	3,481

Sources: Petroconsultants SA, American Petroleum Institute, Paine Webber.

Table 1.10
WORLD – EXCLUDING RUSSIA AND OTHER CIS – ORIGINAL PROVEN OIL RESERVES BY YEAR OF DISCOVERY FROM 1961 TO 1992
By region
(in million barrels)

Year	*Europe*	*Middle East*	*Africa*	*U.S.A./ Canada*	*Latin America*	*Far East China and Oceania*	*Total*
1961	90	11,535	1,241	2,926	713	1	16,506
1962	140	11,175	1,444	2,520	1,163	114	16,556
1963	30	6,744	814	2,611	1,775	701	12,675
1964	185	25,100	1,216	3,307	1,001	1,047	31,856
1965	20	8,100	4,068	3,882	1,143	171	17,384
1966	30	8,000	4,302	3,910	870	253	17,365
1967	60	14,147	5,542	3,108	1,037	1,772	25,666
1968	60	4,990	3,529	2,891	1,143	564	13,177
1969	1,268	6,900	1,409	2,378	1,390	504	13,849
1970	2,580	11,613	636	12,978	1,101	1,811	30,719
1971	2,525	3,34	3,257	2,558	760	2,510	14,950
1972	1,790	2,726	2,284	1,350	1,520	944	10,614
1973	3,290	2,300	1,616	2,355	1,930	1,508	12,999
1974	6,990	3,360	2,104	2,160	2,664	1,744	19,022
1975	2,563	490	1,203	1,496	1,697	2,251	9,700
1976	1,008	8,323	687	1,304	3,999	3,720	19,041
1977	968	500	1,068	1,495	4,727	344	9,102
1978	654	300	418	1,960	19,127	4,357	26,816
1979	4,448	5,355	1,044	1,410	6,937	1,106	20,300
1980	1,258	3,375	1,007	2,970	5,716	1,046	15,372
1981	1,334	656	5,967	2,570	1,568	1,289	13,384
1982	718	2,275	1,886	1,382	1,810	622	8,693
1983	747	3,143	1,386	2,897	3,620	2,180	13,973
1984	2,460	3,303	2,083	3,748	2,386	3,381	17,361
1985	1,542	927	1,561	3,022	3,927	912	11,891
1986	1,119	700	580	1,446	6,408	988	11,241
1987	855	1,380	899	3,240	2,199	1,572	10,145
1988	1,309	1,359	864	2,380	7,879	1,309	15,100
1989	890	3,454	1,173	2,268	2,286	2,216	12,287
1990	1,336	2,070	2,296	2,258	898	801	9,659
1991	1,352	3,403	1,224	940	349	1,162	8,430
1992	1,100	784	927	1,509	1,360	900	6,580

Sources: Petroconsultants SA; American Petroleum Institute, Paine and Webber

Starting from the assumption that the potential of the 450 provinces already explored can be estimated (though some areas remain considerably uncertain), it is reasonable to suppose that exploration will have to be directed towards discovering new energy resources in the provinces where there has been little or no prospecting.

One fact that must not be forgotten, however, is that there is a decline in the rate at which new producing provinces are being opened up. This decline is definitely attributable to the fact that the accessible provinces, those where prospecting costs are reasonable, have on the whole already been explored. To the contrary, some provinces, for example the Arctic ones, with their very considerable logistic and environmental difficulties, will require greater financial resources and new, more advanced technologies.

The new field wildcat (NFW) as an indicator

The number of new field wildcats (NFW) is the best available indicator of the amount of work necessary for the discovery of new oil reserves. In fact the activity previous to drilling of a NFW – that is, geological and geophysical activities, costs much less than exploratory drilling. NFW can be measured counting the actual wells or counting the total metres drilled; we have used the first way, for simplicity and because of the availability of data. Russia and the CIS do not publish figures and are therefore not included. Table 1.9 presents the available data for the NFW drilled from 1961 to 1992 by six regions. This activity brought to light the reserves which have been found in that period of time, which are presented in Table 1.10. The distribution in time of these findings is represented in Figures 1.2–1.8, which give the proven reserves by year of discovery. Figures 1.9–1.15 include the same reserves, but present them as a cumulative value. The two series of figures give an idea of the temporal distribution of discoveries, and therefore will allow us empirically to trace a possible temporal profile of the activity necessary to bring to light the reserves to be found region by region.

The future reserves to be found, region by region, were then spread from 1996 to 2030, according to the empirical profile of new discoveries, taken from the history of the reserves discovered (Figures 1.2–1.15).

Figure 1.2 WORLD (excluding Russia and CIS) - Original estimated proven oil reserves by year of discovery from 1961 to 1992 and forecast to 2030

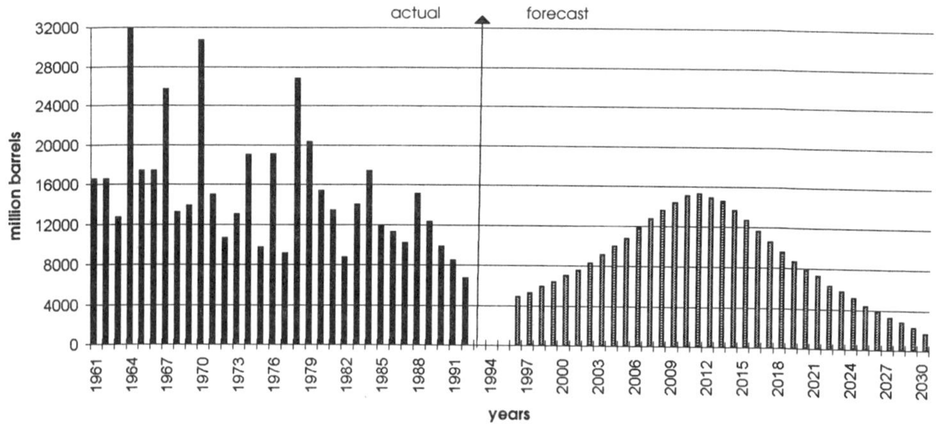

Figure 1.3 EUROPE - Original estimated proven oil reserves by year of discovery from 1961 to 1992 and forecast to 2030

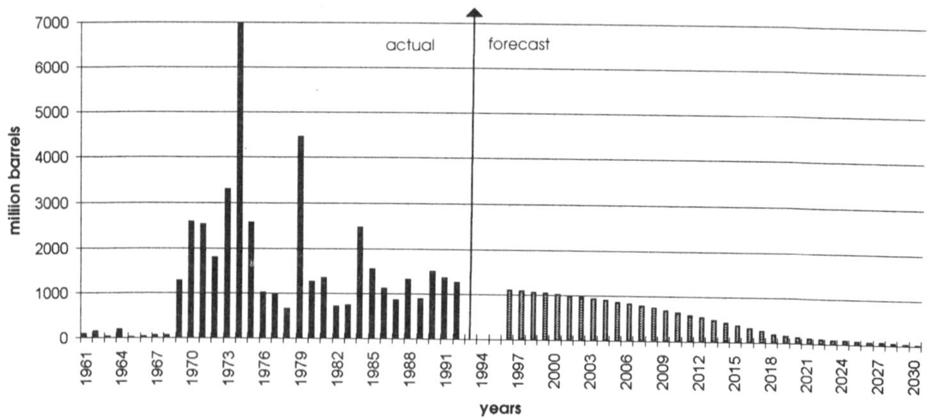

Figure 1.4 MIDDLE EAST - Original estimated proven oil reserves by year of discovery from 1961 to 1992 and forecast to 2030

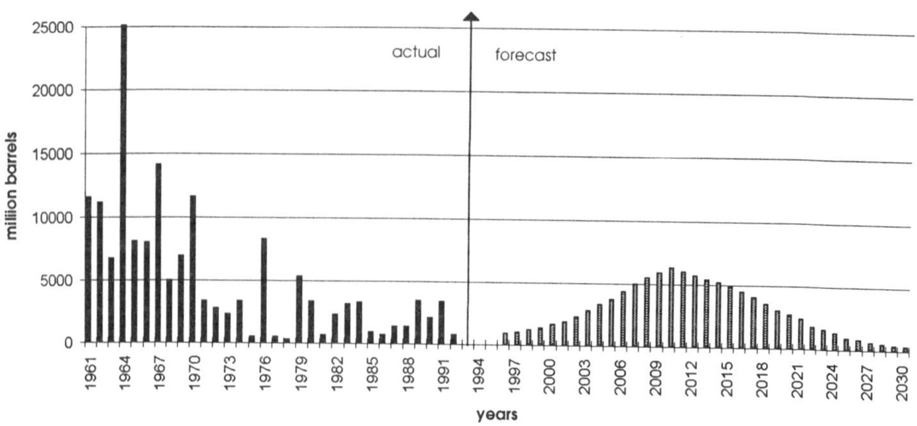

Figure 1.5 AFRICA - Original estimated proven oil reserves by year of discovery from 1961 to 1992 and forecast to 2030

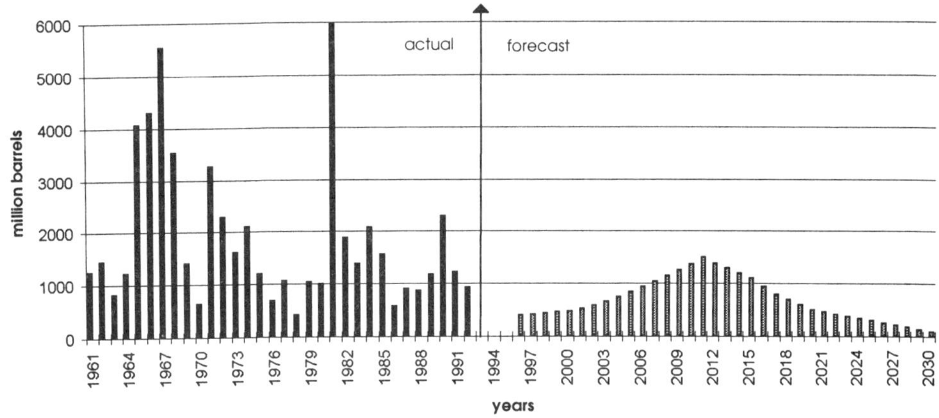

Figure 1.6 USA and CANADA - Original estimated proven oil reserves by year of discovery from 1961 to 1992 and forecast to 2030

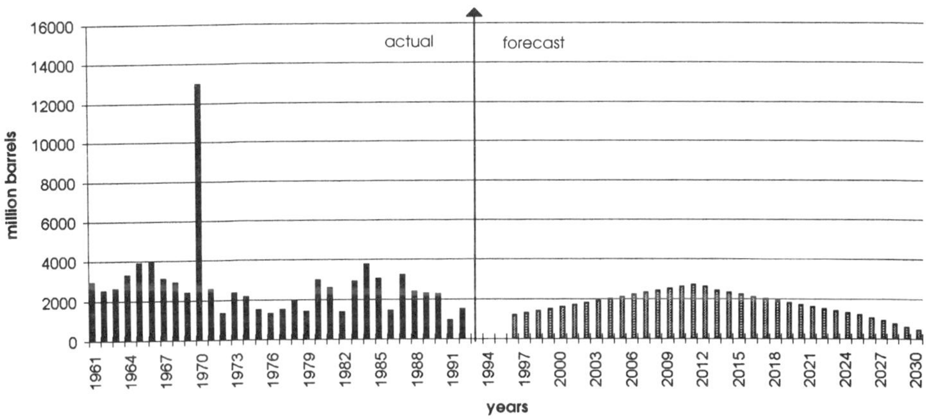

Figure 1.7 LATIN AMERICA - Original estimated proven oil reserves by year of discovery from 1961 to 1992 and forecast to 2030

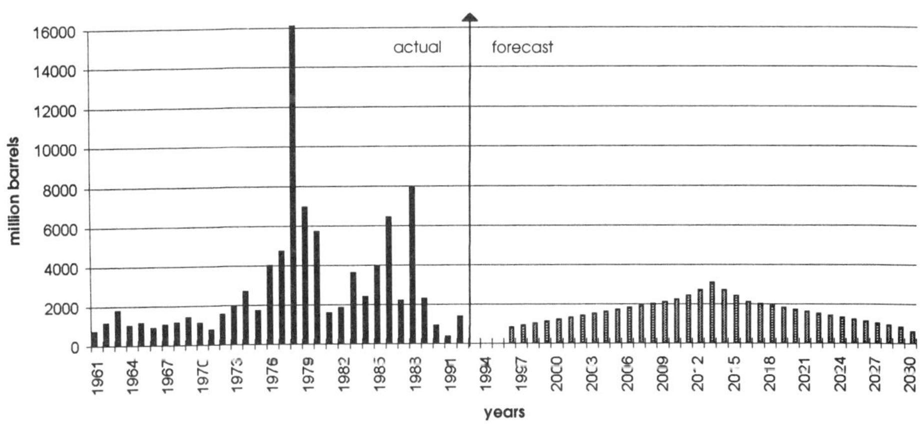

Figure 1.8 FAR EAST - Original estimated proven oil reserves by year of discovery from 1961 to 1992 and forecast to 2030

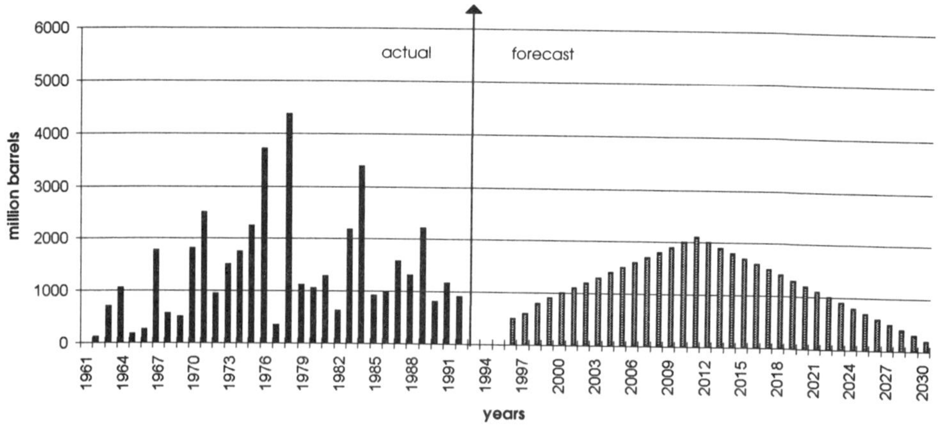

Figure 1.9 WORLD (excluding Russia and CIS) - Original estimated proven oil reserves by year of discovery from 1961 to 1992 and forecast to 2030 (cumulative figures)

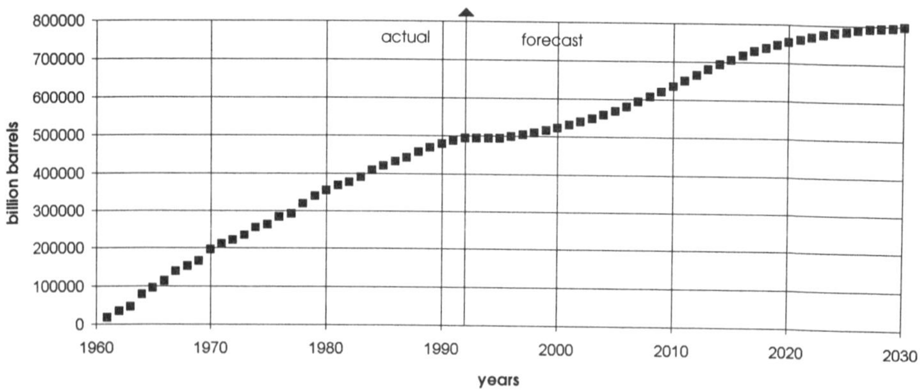

Figure 1.10 EUROPE - Original estimated proven oil reserves by year of discovery from 1961 to 1992 and forecast to 2030 (cumulative figures)

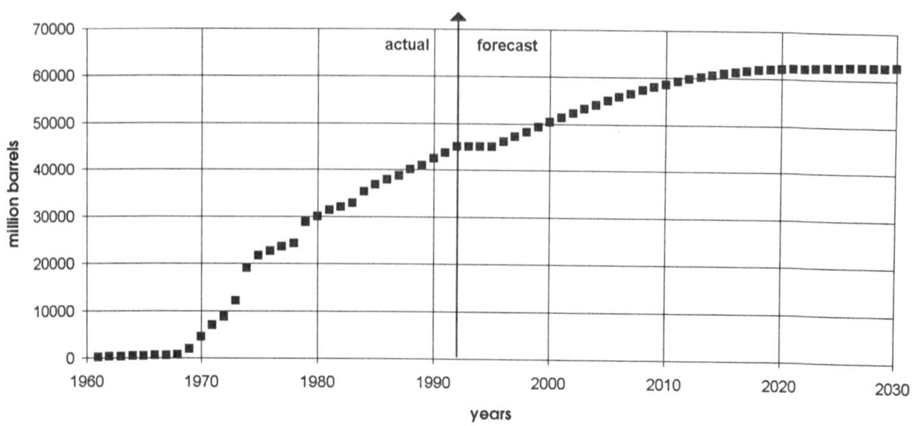

Figure 1.11 MIDDLE EAST - Original estimated proven oil reserves by year of discovery from 1961 to 1992 and forecast to 2030 (cumulative figures)

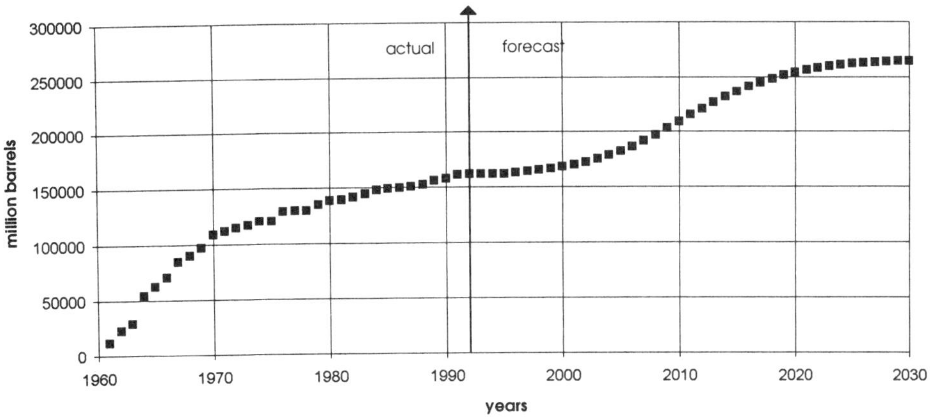

Figure 1.12 AFRICA - Original estimated proven oil reserves by year of discovery from 1961 to 1992 and forecast to 2030 (cumulative figures)

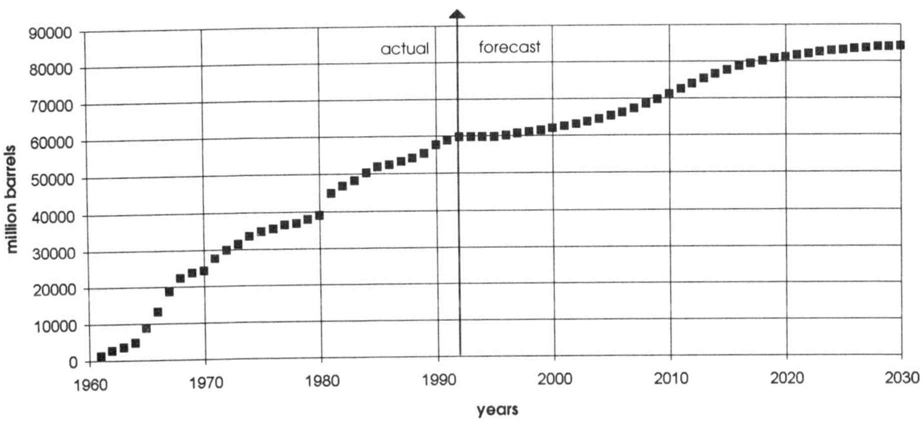

Figure 1.13 USA and CANADA - Original estimated proven oil reserves by year of discovery from 1961 to 1992 and forecast to 2030 (cumulative figures)

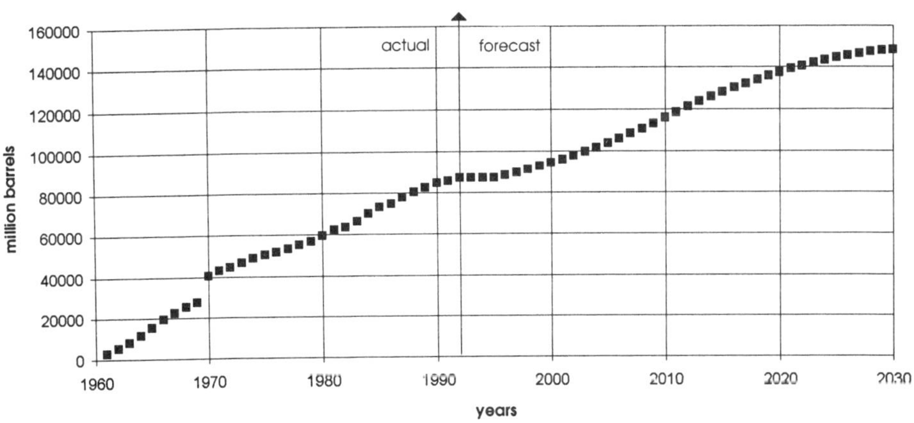

Figure 1.14 LATIN AMERICA - Original estimated proven oil reserves by year of discovery from 1961 to 1992 and forecast to 2030 (cumulative fiigures)

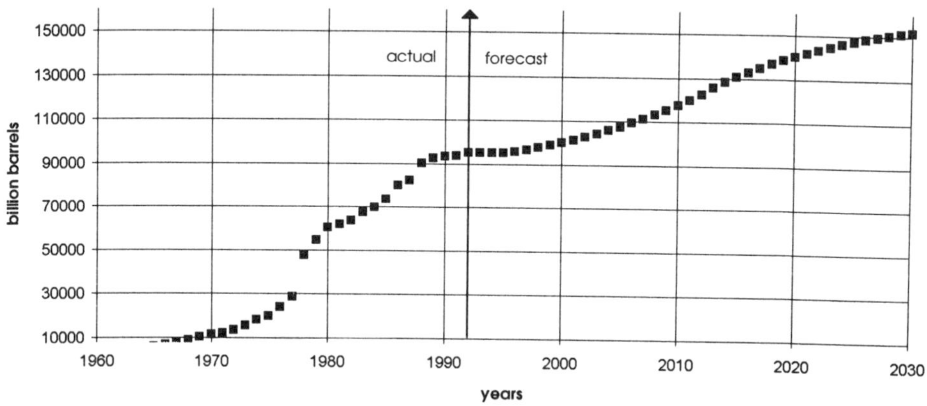

Figure 1.15 FAR EAST - Original estimated proven oil reserves by year of discovery from 1961 to 1992 and forecast to 2030 (cumulative figures)

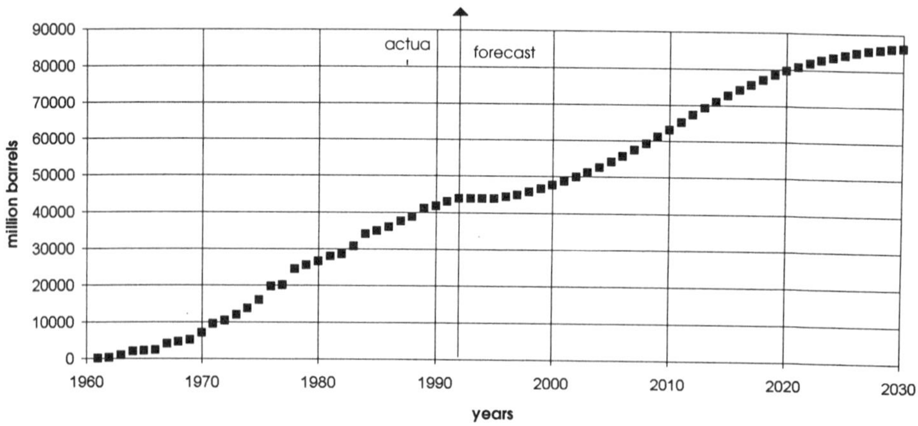

The number of NFWs to be drilled in the future

The trend of the discovery, year by year, of the proven oil reserves found on average per NFW from 1961 to 1992, was calculated (Figures 1.16–1.22), by 7-year mobile averages. We have assumed that, in the future, the reserves found per NFW will follow the past trend. Figures 1.23–1.29 present the number of NFW that would have to be drilled each year up to 2030 in order to discover the estimated potential reserves.

Figure 1.16 WORLD (excluding Russia and CIS) - Original oil reserves found on average per NFW from 1967 to 1992 and forecast to 2030

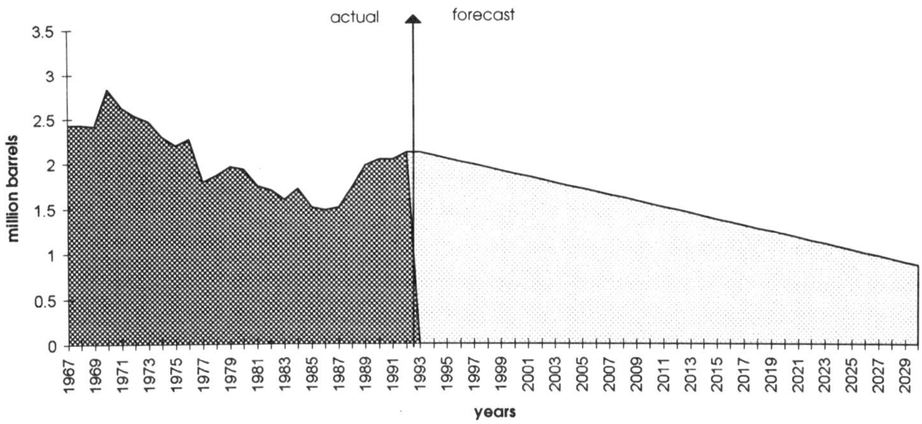

Figure 1.17 EUROPE - Original oil reserves found on average per NFW from 1967 to 1992 and forecast to 2030

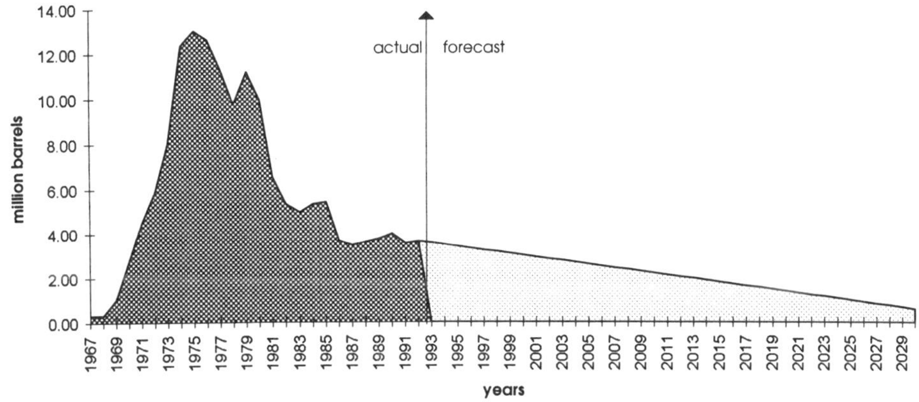

Figure 1.18 MIDDLE EAST - Original oil reserves found on average per NFW from 1967 to 1992 and forecast to 2030

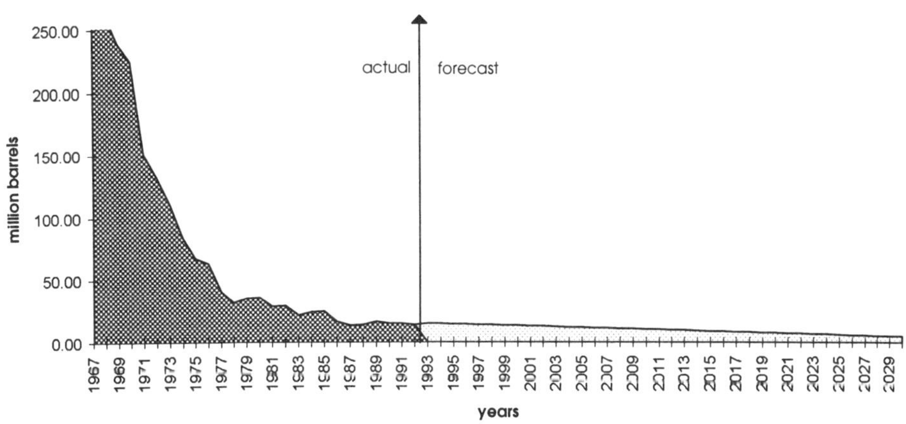

Figure 1.19 AFRICA - Original oil reserves found on average per NFW from 1967 to 1992 and forecast to 2030

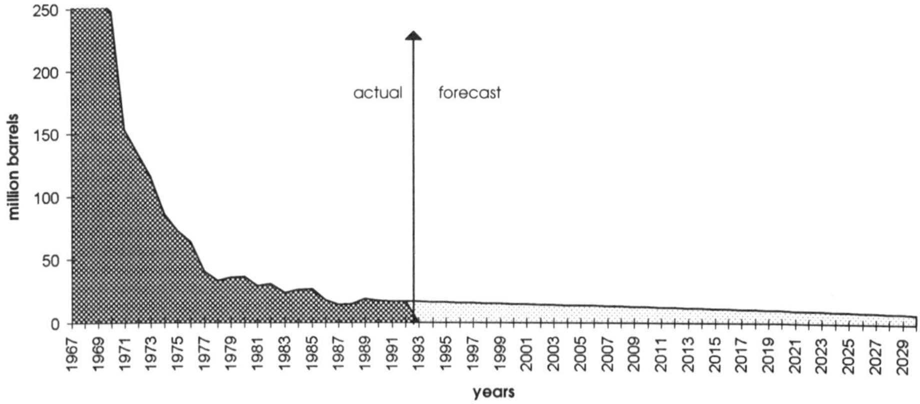

Figure 1.20 USA and CANADA - Original oil reserves found on average per NFW from 1967 to 1992 and forecast to 2030

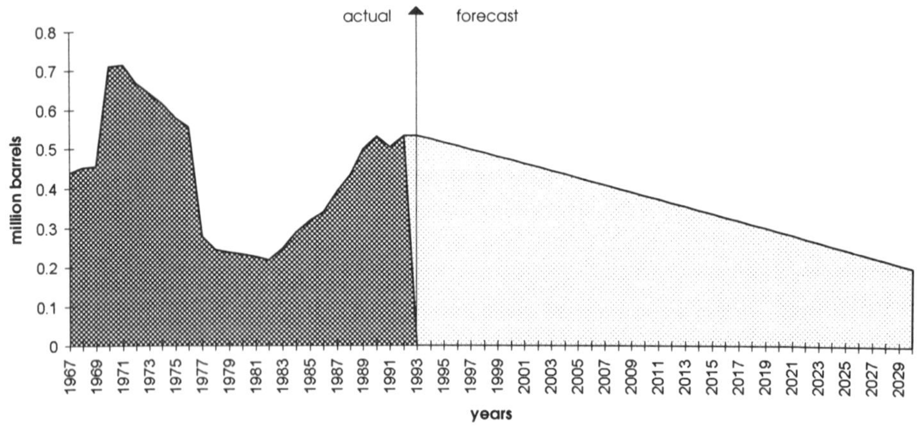

Figure 1.21 LATIN AMERICA - Original oil reserves found on average per NFW from 1967 to 1992 and forecast to 2030

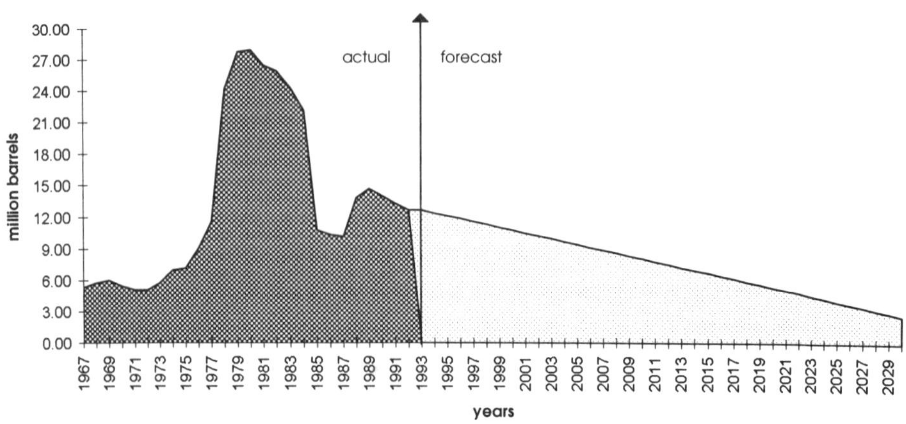

Figure 1.22 FAR EAST - Original oil reserves found on average per NFW from 1967 to 1992 and forecast to 2030

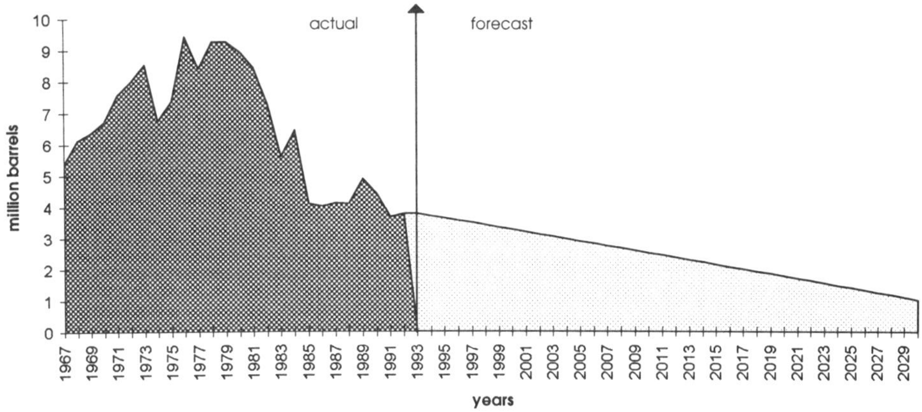

Figure 1.23 WORLD (excluding Russia and CIS) - Number of NFW drilled from 1961 to 1992 and forecast to 2030

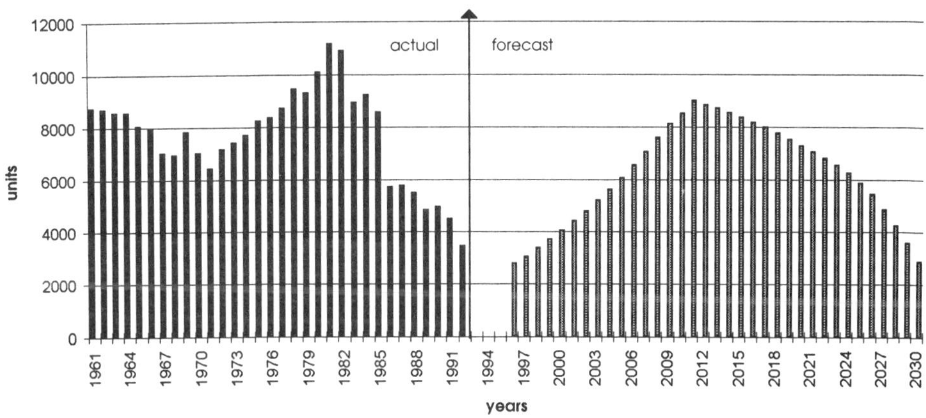

Figure 1.24 EUROPE - Number of NFW drilled from 1961 to 1992 and forecast to 2030

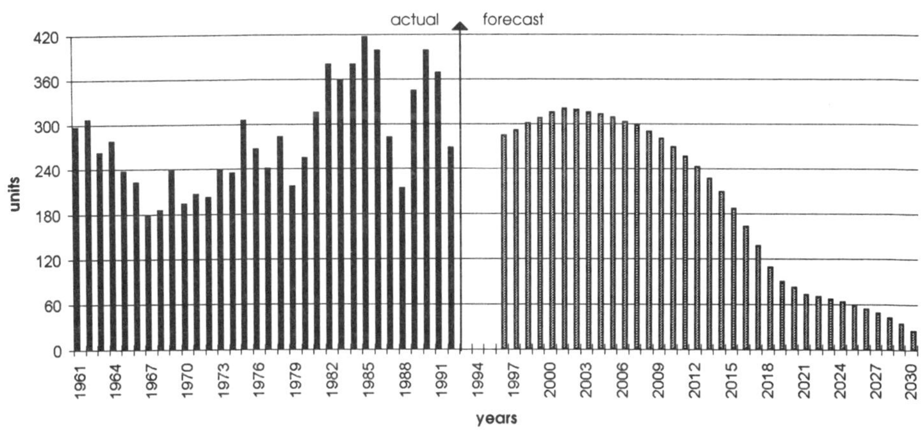

Figure 1.25 MIDDLE EAST - Number of NFW drilled from 1961 to 1992 and forecast to 2030

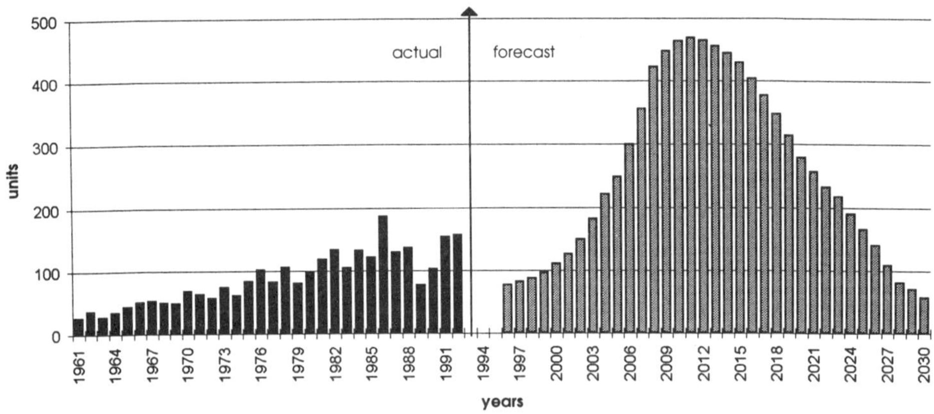

Figure 1.26 AFRICA - Number of NFW drilled from 1961 to 1992 and forecast to 2030

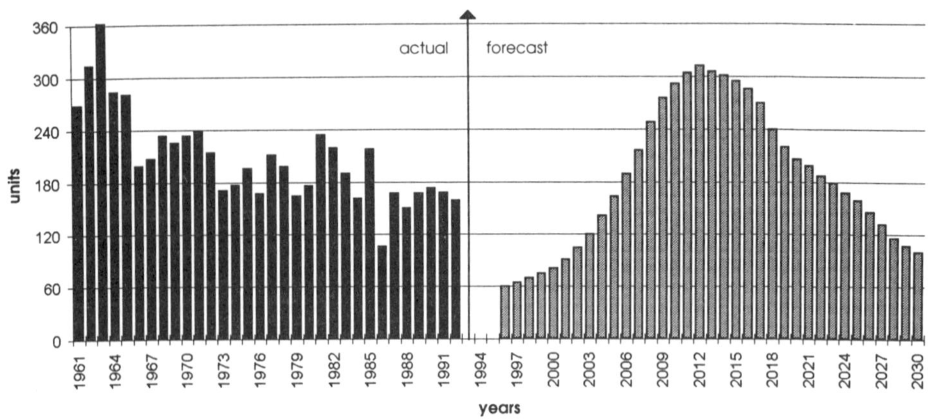

Figure 1.27 USA and CANADA - Number of NFW drilled from 1961 to 1992 and forecast to 2030

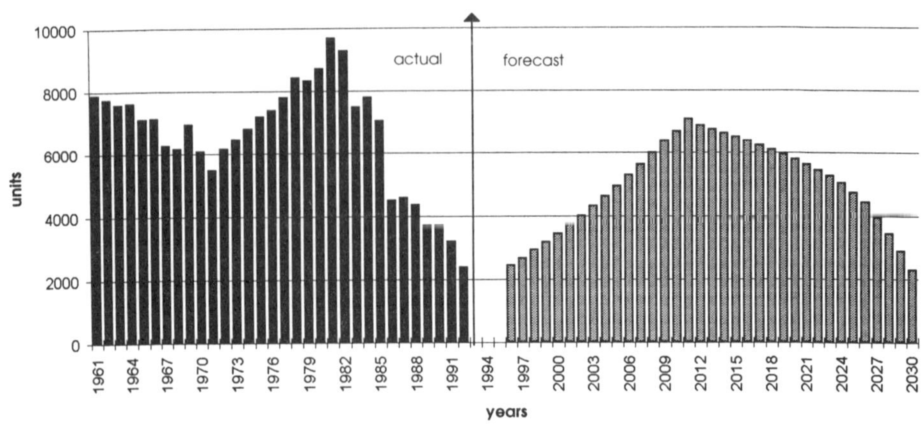

Figure 1.28 LATIN AMERICA - Number of NFW drilled from 1961 to 1992 and forecast to 2030

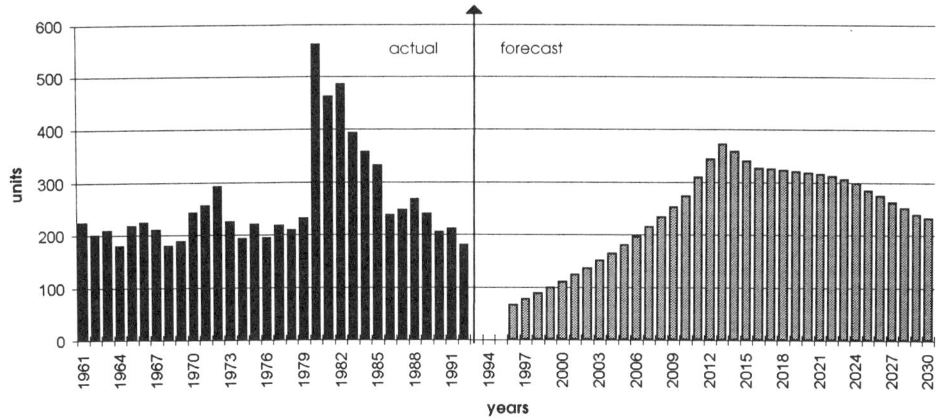

Figure 1.29 FAR EAST - Number of NFW drilled from 1961 to 1992 and forecast to 2030

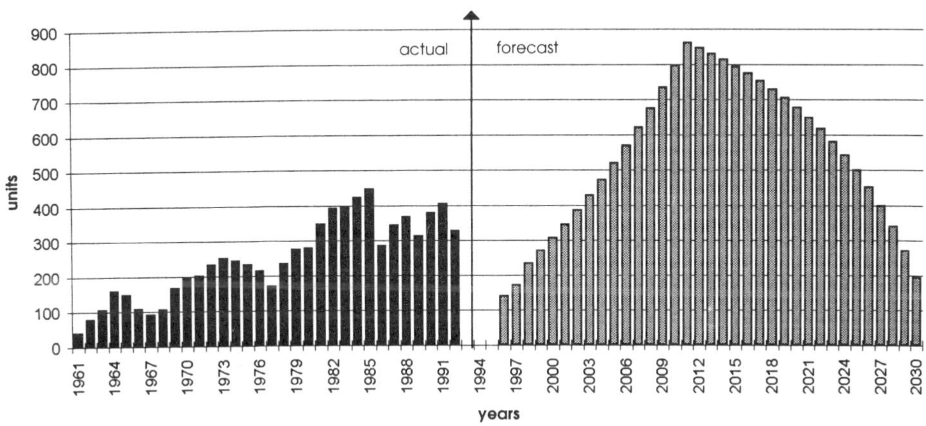

Conclusion

Table 1.11 presents the figures for oil reserves discovered per NFW in the past (line c) and an estimate up to the year 2030 (line e), by region. According to the empirical principle that, in any given area, the biggest fields are the easiest to find, and are discovered first, less oil will be found per NFW from now to the year 2030 than in the past: compare lines *c*, c_1 and *e* of Table 1.11. To find oil in the future will require a much greater exploration effort than ever before, because the quantities discovered per NFW move downwards from the 2.4 million barrels per NFW in 1961–1979 (line *c* of Table 1.11) to 1.7 mb in 1980–1992 (line c_1) and to 1.34 million barrels in 1993–2030 (line *e*).

Table 1.11
ORIGINAL PROVEN OIL RESERVES BY YEAR OF DISCOVERY FROM 1961 TO 1992 AND AN ESTIMATE OF NEW RESERVES TO BE DISCOVERED FROM 1993 TO 2030, BY REGION
(EXCLUDING RUSSIA & OTHER CIS COUNTRIES)

a – Oil reserves discovered from 1961 to 1979 (million barrels)
b – Number of new field wildcats drilled from 1961 to 1979
c – Oil reserves obtained per new field wildcat from 1961 to 1979 (million barrels)
a_1 – Oil reserves discovered from 1980 to 1992
b_1 – Number of new field wildcats drilled from 1980 to 1992
c_1 – Oil reserves obtained per new field wildcat from 1980 to 1992 (million barrels)
d – Forecast of oil reserves to be discovered from 1993 to 2030 (million barrels)
e – Forecast of oil reserves obtainable per new field wildcat from 1993 to 2030 (million barrels)
f – Forecast of cumulative number of new field wildcats to be drilled from 1993 to 2030.

	Year	*Europe*	*Middle East*	*Africa*	*USA and Canada*	*Latin America*	*Far East*	*Total*	*Total without USA and Canada*
a	1961-1979	28,699	144,998	49,053	56,599	54,697	25,422	359,468	302,869
b	1961-1979	4,594	1,104	4,345	134,581	4,110	3,259	151,993	17,412
c	1961-1979	6.2	131.3	11.3	0.42	13.3	7.8	2.4	17.4
a_1	1980-1992	16,233	26,729	21,853	30,630	40,406	18,378	154,229	123,599
b_1	1980-1992	4,198	1,649	2,241	76,461	3,880	4,685	93,114	16,653
c_1	1980-1992	3.9	16.2	9.8	0.40	10.4	3.9	1.7	7.4
d	1993-2030	15,000	99,000	26,000	58,000	56,000	42,000	296,000	238,000
e	1993-2030	2.5	11.1	5.4	0.33	6.9	2.2	1.34	5.0
f	1993-2030	6,000	8,891	4,860	174,045	8,087	19,024	220,907	46,862

The smaller the find per well the greater the number of wells to be drilled: the NFW to be drilled around the world (excluding Russia and other CIS countries) in order to find 296 billion barrels (i.e. 387 billion barrels minus 91 billion barrels from CIS) add up to a remarkable figure, perhaps 220,907 wells (line *f* of Table 1.11) as against the 245,107 NFW drilled (line b plus line b1 of Table 1.11) to find 514 billion barrels up to the year 1992.

This reduction is expected in all areas in the world: however, the absolute amount of oil to be found per NFW in the Middle East is still many times the world's average, because its geological structure still offers the possibility of finding some more supergiants. The size of the reserves that could be found per NFW is subject to wide variations within regions, falling by 30% in the Middle

East, Europe, Africa, and the Far East, and by 20% in Latin America, the USA and Canada. A very large number of NFW will have to be drilled in the US and Canada, namely 174,000 or 78.8% of the total (against 89% in the 1961–1979 period and 82% in 1973–1992); followed by the Far East with 8.6% and by the Middle East with 4.0%.

CHAPTER 2

HYDROCARBON PRODUCTION CAPACITY

Introduction

In the first chapter we presented an estimate of the resources of hydrocarbons in the world, both known and undiscovered, as well as of the exploration efforts necessary for finding them. Basically, we have said that if a certain exploration activity is kept up in the future by the oil industry, a large quantity of hydrocarbon reserves might be found. This would of course require the absence of political, economic and financial impediments to explore the sedimentary basins of the world, the availability of technologies, and the ability to develop every discovery with no external limitations.

However, Chapter 1 did not give any idea of the actual production which could be obtained from such reserves, old and new, in order to satisfy the demand for energy. To do that we may follow three different methodologies, the first based on the ratio between reserves and production; the second on evaluation of production at maximum efficient rate (MER); and the third on a possible trend of hydrocarbon demand.

Reserves to production ratio

The oil industry usually evaluates the duration of proven reserves by the reserves to production ratio (R/P), often called "Static Lifetime of Reserves". This method tells us that the 1993 world oil production (21.53 billion barrels) could be sustained by the proven reserves for 46 years. If we add to the proven reserves the ones which could be found in the future, the number of years reaches 70 without, and 78 with enhanced recovery (see Table 2.1). This ratio is, however, quite misleading, as it implies that the production of 1993 could be sustained uninterruptedly for 46 years, which is clearly not realistic, not only because the demand will not remain flat all the time, but also because oil fields decline during the second half of their life and their production is gradually reduced until they are exhausted.

In the USA it is common practice to compute reserves in a declining field by multiplying last year's production by ten; it will come as no surprise, therefore, to find that US R/P ratio is usually about the same number.

M. Colitti and C. Simeoni, Perspectives of Oil and Gas: The Road to Interdependence, 29–43.

Table 2.1
LIFETIME OF OIL RESERVES (years)

	Static lifetime of reserves (Ratio between reserves and 1993 production)		
Regions	*Proven*	*Proven+ To be discovered+ Re-evaluations and extensions*	*Proven+ To be discovered+ Re-evaluateds and extensions+ Enhanced recoveries*
Europe	10.4	23.2	25.4
Middle East	100.1	121.1	133.1
Africa	27.9	45.9	53.6
USA and Canada	9.4	34.5	41.0
Latin America	47.3	72.7	78.4
Far East	19.1	42.1	46.0
Russia and other CIS	20.1	60.8	72.1
World	46.2	70.3	78.5

Production at maximum efficient rate (MER)

This method estimates the maximum oil production capacity by assuming that each oil field is produced at what the oil industry calls "MER – Maximum Efficient Rate", the maximum rate of production which can be applied to each oil field without causing irreversible harm to the reserves recoverable from that field.

The profile of a field producing at MER is represented in Figure 2.1, which shows how the production increases very fast in the first 4 years, during which time about 25% of the field's total recoverable oil is produced. Then the field goes into a slow decline, and the remaining oil is produced in the following 26 years. According to this depletion scheme – which reflects the maximum production available without reducing the inner energy of the field, and without impairing total recovery – the field reaches its peak in the first 4 years, then reaches its "maturity" and is slowly depleted. This is clearly related to the normal pressure conditions in an oil field, and it is therefore not applicable to gas fields, which have a higher depletion rate and a stronger inner energy.

This method offers the advantage not only of measuring maximum theoretical production, but also of indicating the peak year of any field or aggregation of oil fields; that is, its conditions of "maturity". The MER depletion scheme is, of course, the most convenient for the oil producer, who gets back the money he has

Figure 2.1 Generalized curve of oil production at maximum efficient rate (MER)

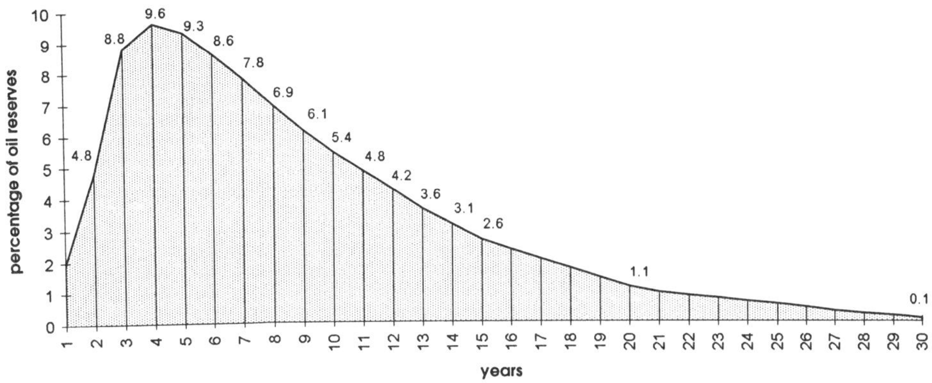

invested in the field in the shortest possible time. This is not always possible, especially for very large fields, whose production at MER would flood the market and push the oil price down to levels which producers would find unacceptable.

The curves of depletion at MER are usually drawn assuming that the development of oil fields immediately follows discovery; that production starts immediately, continues non-stop, and is rapidly raised to the highest level compatible with the maximum ultimate recovery.

The production curves at MER are therefore not a production forecast; they merely indicate the maximum feasible production, as described by Pierre Desprairies of the Institute Français du Pétrole: "Les chiffres qui suivent sont une évaluation de maxima technique de ce que la nature et la technologie permettraient de produire; ils ne constituent pas des prévisions ou des prédictions de production".[1]

The maximum production curves are traced field by field without taking into consideration the fact that the actual oil demand on the market may be lower than the technical capacity to produce.

To answer our question, we have applied the MER depletion scheme to the oil fields of the world taking into consideration not only the reserves actually in existence, but also their re-evaluations, the extra amounts of oil which could be obtained by advanced production methods (enhanced recovery), and, finally, the new reserves that we estimated could be found.

We have worked on a data bank including 1,750 fields of various ages, which have been taken as a representative sample for the 30,000 fields existing in the world. The fields in our sample include 91% of proven reserves, and in 1991 contributed 92% of world oil production.

---

1 10th World Energy Conference, Istanbul 1979

Figure 2.2 WORLD: Oil production at maximum efficient rate (MER)

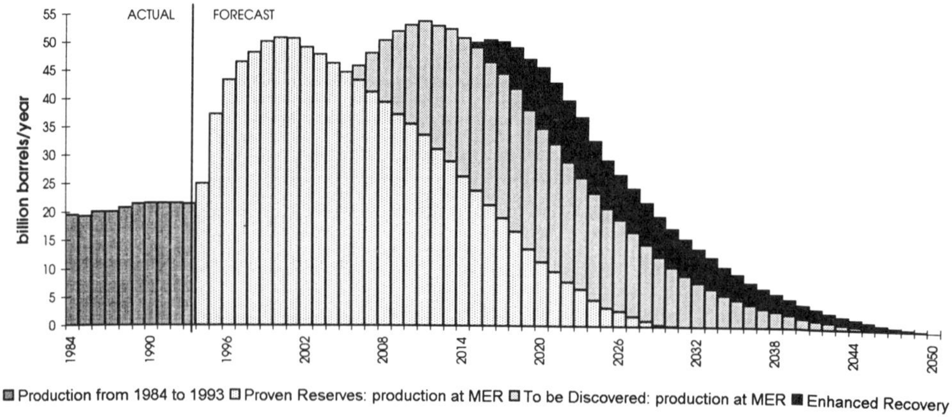

TABLE 2.2
CRUDE OIL IN THE WORLD BY REGION FROM 1993 TO 2030
(billion barrels)

	Resources				
Regions	*Proven at end of 1993* (*a*)	*To be discovered* (*b*)	*Re-evaluations and extension of existing fields* (*c*)	*Enhanced recovery* (*d*)	*Ultimate reserves: proven plus additions up to year 2039* $e = a+b+c+d$
Europe	18.8	15	6	4	43.8
Middle East	662.9	99	40	79	880.9
Africa	62.0	26	14	17	119.0
USA and Canada	28.8	58	19	20	125.8
Latin America	125.0	56	13	15	209.0
Far East and Oceania	44.6	42	12	9	107.6
Russia and other CIS	57.0	91	24	32	204.0
World total	999.1	387	128	176	1690.1

The curve which world oil production would follow if the MER scheme were applied to all fields in the world is represented in Figure 2.2. The trend of production at MER from proven and discoverable reserves, re-evaluations and enhanced recovery (altogether totalling 1,690.1 billion barrels, as shown in Table 2.2) climbs up in the first 19 years, reaching a peak of 54.1 billion barrels in the year 2011 (Table 2.3); then it goes down until the fields are exhausted in the

Table 2.3
PRODUCTION PEAKS AT MER OF PROVEN AND TO-BE-FOUND RESERVES
(billion barrels)

			Production peaks at MER relating to:					
	Production peaks		*Proven reserves at end 1993*		*Proven reserves+ To-be-discovered+ Re-evaluations and extensions*		*Proven reserves+ To-be-discovered+ Re-evaluations and extensions+ Enhanced recoveries*	
Regions	*In the period 1984–1993*	*In 1993*	*Year*	*Value*	*Year*	*Value*	*Year*	*Value*
Europe	1.80	1.80	1994	1.94	2013	1.52	2016	1.48
Middle East	6.60	6.60	2001	32.04	2008	30.18	2016	26.98
Africa	2.29	2.22	2000	3.45	2011	3.42	2018	3.90
USA and Canada	3.76	3.07	1994	3.02	2013	3.56	2019	4.14
Latin America	2.66	2.66	2001	6.16	2012	7.69	2015	7.08
Far East	2.34	2.34	1997	3.11	2011	3.51	2017	3.15
Russia and other CIS	4.56	2.81	1997	3.65	2013	5.46	2019	5.71
World total	21.59	21.53	2000	51.82	2011	54.12	2016	51.74

Table 2.4
PROVEN AND TO-BE-FOUND OIL RESERVES
(number of years in which production at MER is greater than 1993 production)

	Reserves		
Regions	*Proven*	*Proven+ To-be-discovered+ Re-evaluations and extensions*	*Proven+ To-be-discovered+ Re-evaluations and extensions+ Enhanced recoveries*
Europe	2	0	0
Middle East	29	33	36
Africa	16	28	34
USA and Canada	0	12	19
Latin America	22	31	33
Far East	9	21	26
Russia and other CIS	10	37	43
World	23	31	35

second half of the 2040s. If we limit ourselves to the proven reserves, maximum production at MER would occur in the year 2000 (see Table 2.3) with 51.8 billion barrels (144 million barrels/day).

We can conclude that world oil production has not yet reached its all-time peak: the oil industry has not yet reached its "maturity", and it is not yet on the point of starting its decline.

At world level the difference between what would be produced if all the fields were exploited at MER and historical production is of course very large (Figure 2.2).

If we compare the production at MER with that of 1993 (Table 2.4) and with the historical maximum recorded in 1990 (21.59 billion barrels) we find that it exceeds the amount recorded in 1993 for the first 35 years (from 1994 to 2028), and the maximum for the first 31. This means that the production level recorded in 1993 could be kept without interruption for 35 (Table 2.4), or 31 years. In 28 of these 35 years production at MER is more than double the 1993 figure. After these years the MER production starts dropping in 2030, to cease in 2049.

Our data bank has allowed us to draw similar curves for the various areas in which the oil industry divides the world. Figures 2.3–2.9 show that the MER method is not purely theoretical. The fields of the producer–consumer countries,

Figure 2.3 EUROPE: Oil production at maximum efficient rate (MER)

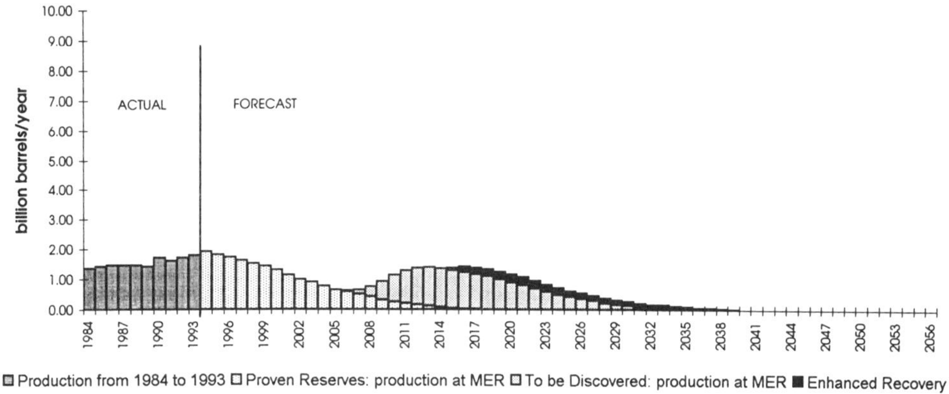

Figure 2.4 MIDDLE EAST: Oil production at maximum efficient rate (MER)

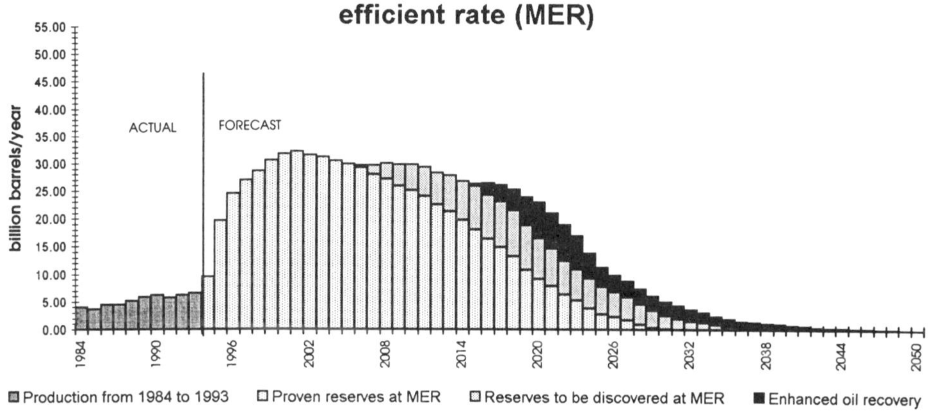

Figure 2.5 AFRICA: Oil production at maximum efficient rate (MER)

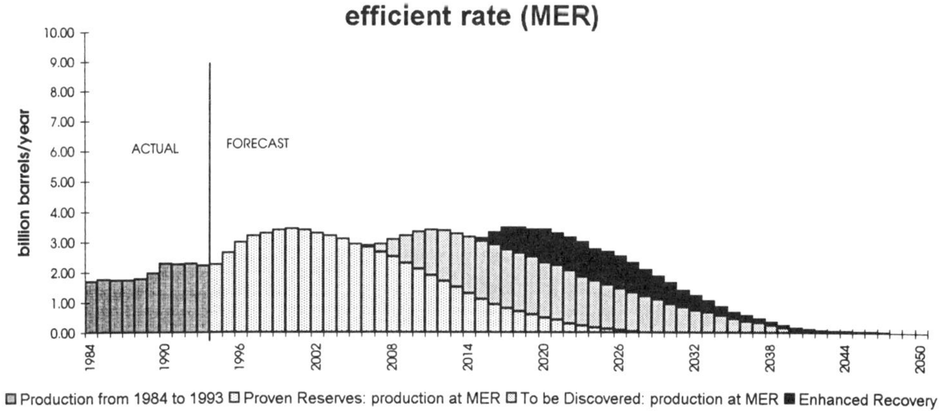

Figure 2.6 USA and CANADA: Oil production at maximum efficient rate (MER)

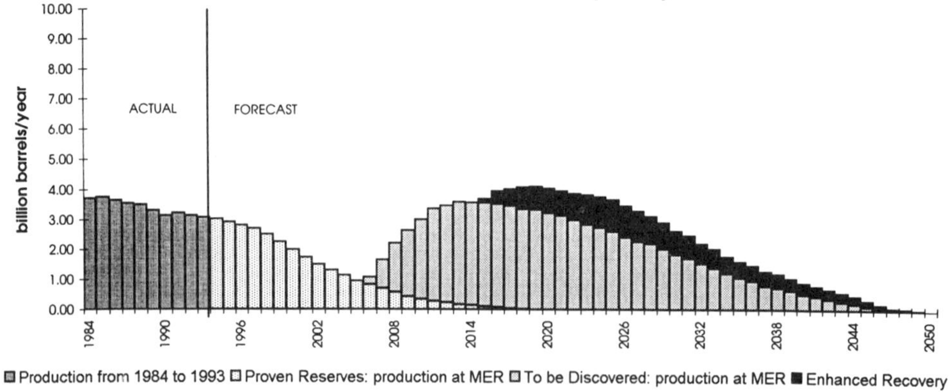

Figure 2.7 LATIN AMERICA: Oil production at maximum efficient rate (MER)

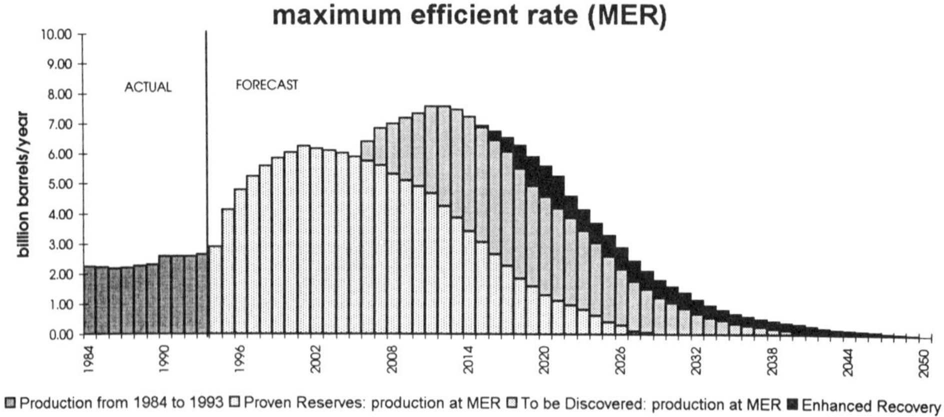

Figure 2.8 FAR EAST: Oil production at maximum efficient rate (MER)

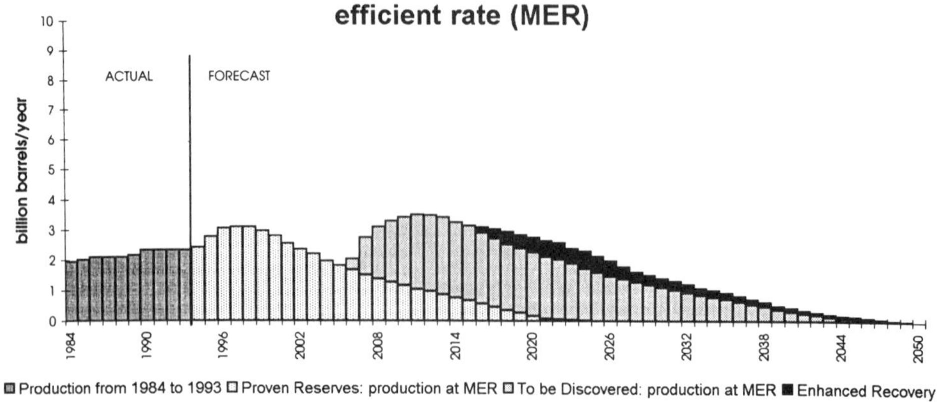

Figure 2.9 RUSSIA and CIS: Oil production at maximum efficient rate (MER)

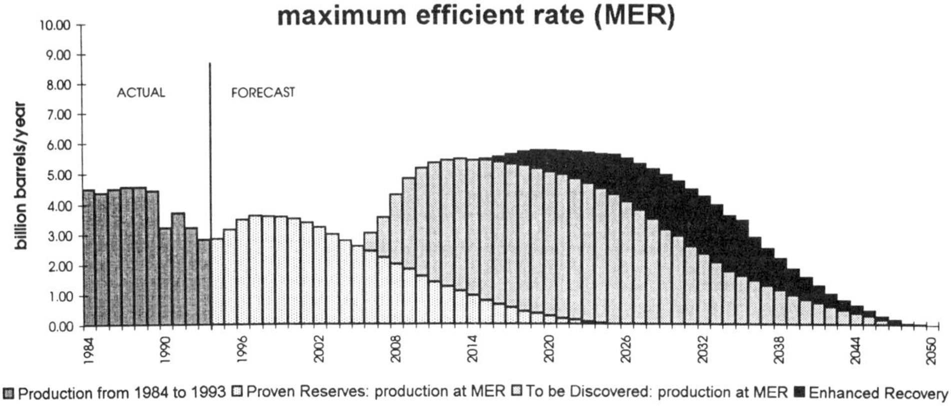

and particularly the USA, Canada and Europe, produce at MER or close to it. The same can be said of a large exporter, the former USSR. To the contrary, the large fields of the Middle East, and – at a lower level – those of Latin America, have never produced at a rate at which their output would have exceeded the demand for their oil.

In Europe (Figure 2.3), an area of strong oil demand, the actual production, although very near to the production at MER, reached its maximum in 1994 with 1.94 billion barrels. The production of 1.80 billion barrels recorded in 1993 could be kept up in 1994 and 1995, after which it would decrease, the reserves becoming exhausted around 2039.

In the Middle East (Figure 2.4) the difference between production at MER and actual production is very large: the fields there are exploited well below their potential. The area would reach this peak in 2001 with 32.0 billion barrels with the proven reserves only, and in 2008 with 30.2 billion barrels adding the future ones. The production level of 6.6 billion barrels recorded in 1993 could be kept up without interruption for 29 years with the help of proven reserves, and for 36 years with the addition of reserves to be found.

In Africa (Figure 2.5) the difference between production at MER and actual production is not excessive: in that area output has been kept high compared with potential. Oil production has not yet reached its maximum level: peak production at MER of the proven reserves will occur in the year 2000 with 3.45 billion barrels, and in 2018 with 3.90 billion barrels with the help of reserves yet to be found.

The production level of 2.22 billion barrels recorded in 1993 could be kept up without interruption for 16 years with the help of proven reserves (then diminishing from the year 2010 and becoming exhausted in the year 2027), and for 34 years with the contribution of the reserves to be found.

In the USA and Canada (Figure 2.6) production at MER and actual production are very near. The maximum production level has already occurred in 1985 with 3.8 billion barrels; it would occur in 2019 with 4.14 billion barrels with the contribution of reserves to be found.

The production level of 3.07 billion barrels recorded in 1993 cannot be retained with the help of proven reserves, but could be kept for 19 years with the contribution of reserves to be found.

In Latin America (Figure 2.7), the difference between production at MER and actual production is very large, similar – although at a lower level – to that of the Middle East. Oil production there has not reached its peak: with the proven reserves the area will peak in the year 2001 with 6.16 billion barrels, and in the year 2012 with 7.69 billion barrels with the contribution of the reserves to be discovered.

The level of 2.66 billion barrels produced in 1993 could be kept up uninterruptedly for 22 years with the help of proven reserves, and for 33 years with the addition of reserves to be found.

In the Far East (see Figure 2.8) oil production is lower than its potential maximum. It will peak in the year 1997 with 3.11 billion barrels from proven reserves only. Adding reserves yet to be found, the peak would be reached in 2011 with 3.5 billion barrels. The level of 2.34 billion barrels recorded in 1993 could be kept up without interruption with 9 years on proven reserves alone, which would decline and become exhausted around 2023. Adding reserves yet to be found, the level of 1993 could be kept for 26 years.

In Russia and other CIS countries (see Figure 2.9) the difference between production at MER and actual production is minimal. Maximum production has not yet been reached: peak production at MER from proven reserves would occur in 1997 with 3.65 billion barrels; and in 2019 with 5.71 billion barrels with the help of reserves to be discovered by the year 2006.

The 1993 production level of 2.81 billion barrels could be kept up without interruption for over 10 years from proven reserves alone, which would peak in the year 2006 and be exhausted by 2024; and for 43 years with the help of reserves yet to be found.

The level of crude oil (and of natural gas) production according to a demand trend

The second method has produced many more answers than the first, but it is still not very satisfactory. It gives us an estimate of the production capacity of the world oil industry, but at the cost of applying criteria which are not fully realistic. The large fields of Middle East and Latin America have never been exploited at MER, and a change in this situation is not in sight. Moreover, the second method does not allow us to compare production capacity with what the world might need in the future.

Figure 2.10 WORLD - Oil and natural gas production from 1985 to 1993 and forecast to 2085 (up to the exhaustion of the reserves). Cumulative figures

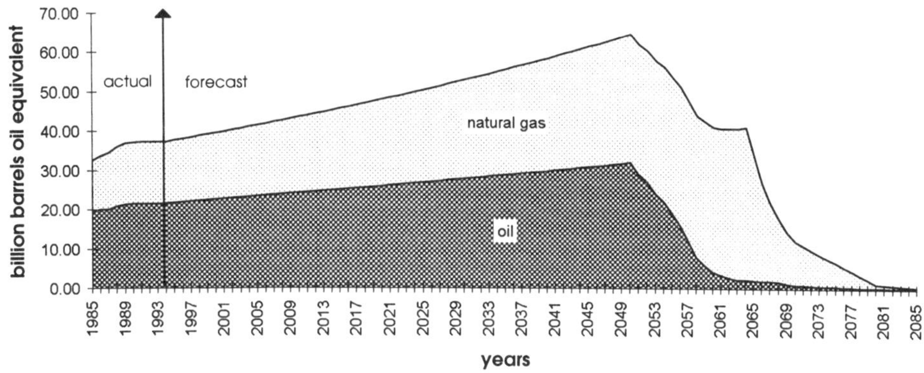

Finally, the comparison between production capacity and possible future demand has to consider natural gas. Gas reserves, as we have seen in the previous chapter, are of the same order of magnitude as those of oil, and gas now supplies about 20% of world energy, being well suited to almost the same uses as oil products (excluding only the production of petrochemical intermediates, ethylene, propylene, c4 and pygas).

The problem here is that the present study does not attempt to forecast demand, but only to investigate how much oil is on the ground, and what should be done to find it. However, in order to answer the obvious question of how long the hydrocarbon will last, let us assume, as an exercise only, that the demand for crude oil in the world will increase in the future at a rate of 0.7% per year (see Figure 2.10) a figure which does not result from a market assessment but it is loosely taken from the trend of past demand.

In 1993 world oil production was 21.5 billion barrels (i.e. 59.7 mb/d). Applying the rate of 0.7% per year we increase yearly production up to the year 2050 which, in this hypothesis, is the "year of maturity" of the oil industry, in which the majority of oil fields are on the way to exhaustion, and the new ones are not large enough to maintain the production level. In the year 2050 oil production peaks at 32.0 billion barrels (88.8 mb/d). The cumulative production for all the years from 1994 to 2050 is 1,317 billion barrels. After that year, reserves can no longer support growth, and production begins to fall .

The profile of world oil production after 2050 has been drawn empirically following the MER principle that production declines gradually. Production will therefore be 19 billion barrels (53 mb/d) in the year 2055; 4 billion barrels (11 mb/d) in 2060; 2.1 billion barrels (5.8 mb/d) in 2065; 1 billion barrels (2.8 mb/d) in 2070, reaching almost zero in the year 2085.

The same methodology has been applied to natural gas, with two changes: first,

Figure 2.11 WORLD - Oil and natural gas production from 1985 to 1993 and forecast to 2085 (up to the exhaustion of the reserves). Cumulative figures

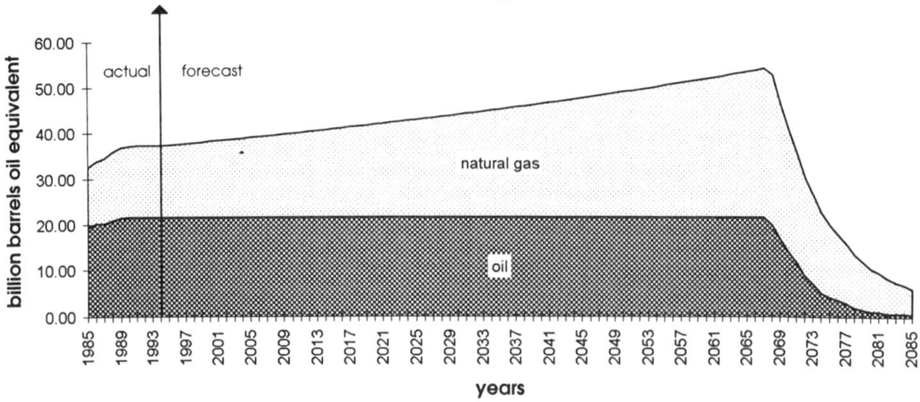

the projected yearly rate of production increase has been set at 1.5% rather than 0.7%; second, due to the geological characteristics of gas fields, production starts declining at a later date, that is, in 2065, but with a much sharper decline.

Natural gas production was 15.7 million barrels of oil equivalent (2,385 billion cubic metres) in 1993. The annual increase of 1.5% makes it peak at 39 million barrels equivalent (6,490 billion cubic metres/year) in the year 2064, with a cumulative production of 1,801 million barrels equivalent (274 trillion cubic metres). After 2064, world gas reserves would not support such a production, which decreases rapidly. The last reserves of natural gas are produced in the year 2085.

A second exercise could be performed by choosing a more conservative future profile: suppose that oil production remains unchanged from 1994 onwards, while gas increases by 1.0% a year (see Figure 2.11). These changes do not fundamentally alter our results: oil production peaks in the year 2067, that is, 17 years later than in the previous exercise; gas peaks in 2068 as compared with 2064, that is, only 4 years later. The decline period will bring us to the year 2085 for the exhaustion of oil and gas reserves.

Conclusion

The numbers we have shown above are, in our opinion, quite self-explanatory. Only a few words are therefore necessary as a conclusion. If we exclude the first method, which is only a rough-and-ready rule of thumb, we can see that the second and the third methods give us a fairly consistent answer, actually two consistent answers.

The oil industry has not yet reached its peak, and it is not entering the era of decline, which may start as far in the future as the year 2016. Only two areas have already entered in the last phase of oil production, namely the USA and Europe; that is, the first ever area of systematic oil extraction, and a marginal area where relevant oil reserves were not found until the end of the 1960s. The other areas still are relatively "young", and at least one very large one, the Middle East, is very young.

A rate of increase of demand not very different from the one of the most recent years would prolong the life of hydrocarbon reserves for almost another century, and similar results were obtained by projecting a flat 1993 production. We can therefore conclude that the present oil industry seems to have a relatively long time horizon, stretching into the future for almost a century.

Are those answers credible? The methodology employed has been not only quite painstaking, but also quite conservative. For example, we have not included in our assumptions any important technological development which would allow us to "see" and to operate in areas not yet reachable today. However, we know that technology improves every day, and important progress is actually quite probable. Our conservative attitude towards technological developments has also been applied to enhanced recovery. We have included only those technologies which seem to be applicable according to a quite tight price assumption, that is, to a price of oil not beyond $21 per barrel, not very far from the present price. So we can conclude that the answers are relatively credible. The margin of error is of course very large, both because of the amplitude and variety of the industry, and the inductive character of what we can know about oil reserves.

It may be useful to point out what our answers do not say. First, they are not forecasts. We did not intend to predict that those quantities of new oil reserves will actually be found. We only said that, if a certain quantified level of activity in petroleum exploration is maintained, the results will very probably be what we have anticipated. Second, we did not range the oil and gas reserves according to their discovery or production costs; that is, we did not introduce in our argument an estimate of the effect of any given oil price to the reserves put into production.

Within the limits which we have stated, we are quite confident that the answers are credible, and would endorse the conclusion that the hydrocarbon era is not yet on the way out, and will last for at least another century. There are therefore all possible reasons for the producers of hydrocarbons to adopt a long-term view of their production and their market, unless their own reserves position turns out to be far below the world average.

Methodology followed in assessing world production levels at the maximum efficient rate (MER)

To assess oil production at MER from the proven reserves at 31 December 1992, data should be collected for each of the about 30,000 oil fields known in the world, which is clearly impossible. We have used a sample of 1,750 oil fields, the largest ones, which in 1993 accounted for 91% of world oil proven reserves and contributed 92% of world oil production.

Our data bank lists such fields by proven reserves; location; country; onshore or offshore; initial year of production; production per year, and cumulated up to the year 1992; proven reserves.

The other remaining oil fields have been grouped country by country under the item "other fields".

Total reserves to be found in the period 1996–2030 amount to 691 billion barrels recoverable. 56% of them will come from newly discovered fields; 19% from re-evaluations or extensions of reserves already known; 25% from enhanced recovery.

The assumption was made that enhanced recovery will bring total production up to an average of 40% of the oil in place. Although the majority of the fields will not justify the application of such techniques, they will certainly be applied to a limited number of giant fields, whose production will go beyond the 40% mark.

The discovery of new reserves and the re-evaluations and extensions have been distributed in the years after the year 2006 according to the trend of past discoveries.

We define: "The productive regime at the MER is the rate at which a well or a pool produces most efficiently. This abbreviation also means maximum economic recovery, which is generally considered to be the result from the highest rate of production without waste, waste being the unreasonable use of reservoir energy to produce oil".

Of course, MER is the production profile which allows the oil company to get back its investment in the shortest possible time.

The MER of most pools (i.e. fields) ranges between 3% and 9% of the recoverable reserve per year: the ultimate life of the field is assumed at 12–33 years.

The annual rate of production at the MER (Figure 2.1) will increase rapidly in the first few years after starting production up to a maximum of about 10% in the 4th year, following which there is a progressive drop until exhaustion. By the 7th year 50% of the extractable reserves would already have been produced; 80% by the end of the 12th year and 90% by the end of the 16th year. After the 20th year production is very low, and it is continued after that year only in particularly favourable geopolitical situations.

For example, in 1992 in the United States the "stripper wells" accounted for 74% of total production wells, but only 13% in terms of US oil production.

In any case an intermediate time period must be allowed between the discovery of the field and its coming into production, to ascertain, prepare and implement the development plan (execution of basic works, wells, facilities, etc.) at the level required from the MER production. This period is extremely variable from basin to basin and from country to country, and is linked essentially to technical, environmental and logistic problems. An average time-value of 4 years has been taken for onshore and 7 years for offshore production.

Estimating the MER production of fields already in production at 31 December 1993 is complicated by the fact that past production has been influenced by non-technical factors (related to demand, prices, or production policies) which have normally kept it below the potential maximum.

Hence assessment of the maximum residual annual production for many of the fields in production requires a thorough knowledge of the conditions of the field, which is not always available. Many fields in production since before the 1970s, especially in the United States, Venezuela and Algeria, are in decline, with the exception of a few supergiants; the fields that came into production between 1970 and 1980 are mainly in the high phase of their cycle; while the majority of the Middle Eastern fields are in the initial phase.

For each of the fields in production, the available figures of the proven reserves, and annual and cumulative production were used to calculate how much of proven reserves has already been produced. Such a percentage was then positioned on the curve that describes the entire production cycle of the field, i.e. the curve of the cumulative rates of production at MER.

For example: a field with an ultimate recovery of 52 billion barrels of oil, and which has already produced 17.9 billion barrels (34.4% of total reserves) will be classified as being in the 5th year of the MER production, the year in which the reserves are depleted by 34%. Figure 2.1 also shows that in the 6th year of production, which in our case corresponds to the year 1994, the annual rate of MER production is equal to 8.6% of the reserves. If the application of the rate for the 6th year gives a production which represents an increase of less than 25%, the field is supposed to have been depleted at MER. If, to the contrary, by applying such a rate we obtain a level of production higher by more than 25% than the previous years, we conclude that the field has been depleted more slowly than MER, possibly due to market or technical constraints. We therefore repeat the production estimate applying the 25% rate, until we connect with the MER levels.

The sum total of the production figures at MER from the proven reserves for each year represents the maximum production capacity.

The aggregate of the annual production at MER of the proven reserves of the fields existing in each geographical areas, or in aggregates of geographical area, represents maximum crude oil production level in time.

Adding the maximum production levels relating to the reserves still discoverable between 1994 and 2030, to the re-evaluation and extensions and enhanced recovery reserves, we obtain the curve of maximum production capacity.

PART II

THE DEVELOPMENT OF THE OIL INDUSTRY

CHAPTER 3

A PICTURE OF THE OIL BUSINESS TODAY

Introduction

The previous chapters have given an answer to the all-important questions: How much oil and gas is there? How long is the hydrocarbon era going to last? The answer has been: there is plenty, but we will have to look for it; if we find it, develop and produce it, it will last for almost another century. The problem has therefore been moved from geology to industry; and industry means primarily a market from which the industry gets its signals, its prices and its cash flow.

Part Two of this book will therefore discuss at length the oil market, starting with a presentation of its present configuration, and explaining in some detail how this structure came to be. It will by necessity concentrate on oil, the older of the hydrocarbons, although some aspects of gas will also be discussed.

The oil market as a "spaghetti junction"

The word "market" usually conveys the image of some place – a square, a large room, a shop – where people buy and sell some commodity; or a number of such places, connected, perhaps, by an information network.

Oil is a fluid, and it moves in large flows out of the oil fields into ships and pipelines, storage tanks, and finally into refineries; from there, oil products flow into more tanks, pipelines, ships and, finally, to consumers' tanks, engines or burners.

Along this route, crude oil and oil products are bought and sold by many buyers and sellers; and many different prices are asked and paid for them.

The words "oil market" therefore convey rather the image of a bunch of flows, a maze, or a "spaghetti junction": Figure 3.1 presents this in a simplified, qualitative way, and makes it look like a game of snakes and ladders. Oil moves along its flow-lines because of the decisions of a small number of important actors, namely the countries which produce and export oil, and the oil companies.

The flow starts with the production of crude from the existing reserves, either by a producing country or an oil company; crude oil is then sold by the countries to the companies, and the latter interchange some of it among themselves directly or through the spot market; part of it goes into stocks; finally, the bulk of the

M. Colitti and C. Simeoni, Perspectives of Oil and Gas: The Road to Interdependence, 47–63.

crude goes through the refineries to be transformed into products and sent to the second maze, eventually to reach the consumers. Figure 3.1 does not, however, contain information about the quantities being produced and exchanged.

The graphs and tables following present a quantitative view of the oil market limited to the first part, the crude oil, which is the main theme of the second part of this book. These graphs are drawn up in two different ways. In the first group the emphasis is on the flow and the connections between the actors, and the graphs look like a spaghetti junction; in the second, the space covered on the page is related to the volumes traded, and the graphs look like horizontal missiles. The two are presented one after another, to give a complete view of the phenomenon. Finally, a numerical table gives all the figures for each of the 4 years.

The data are essentially based on published information, taken from specialised magazines, daily news and annual reports of oil companies. On these data we estimated the figures in the graphs and tables. The choice of the 4 years is also of relevance.

The structure of the oil market we present is, of course, the result of a complex historical change, to which we will dedicate the whole second part of this book.

Suffice now to say that 1980 is the first year of the structure as we know it today, i.e. the first year after 1979, the year of the sudden jump in oil prices which took the name of second oil shock; and that 1992 is the last year for which the information required is available. The two years in between, 1983 and 1986, are first a "normal" year and second the year in which the production and price policies of the producing countries were drastically changed (1986). We are perhaps anticipating too much of the content of the book, which must be developed in a certain order.

The actors and the figures

Let us consider Figures 3.2–3.9. We see three major actors – two groups of countries and the oil companies – exchanging oil along the line which starts with the oil wells and ends into the refineries.

The group at the top of the graphs is made up of Russia plus the countries of the former USSR tied by the CIS treaty. Russia itself is the largest oil producer of the world, and has exported oil almost since the inception of the oil industry. For the best part of this century its oil industry has been strictly centralised, without any private and foreign company. This system is perhaps now on the way to change, but the figures for 1992 still reflect the old structure.

The second group – the exporting countries – is made up of oil exporters: the 12 members of OPEC (the Organisation of Petroleum Exporting Countries), and the others, e.g. Mexico and Egypt. Some of these countries – Saudi Arabia (the largest producer in OPEC), Iran, Iraq, Venezuela, Mexico, etc. – produce crude

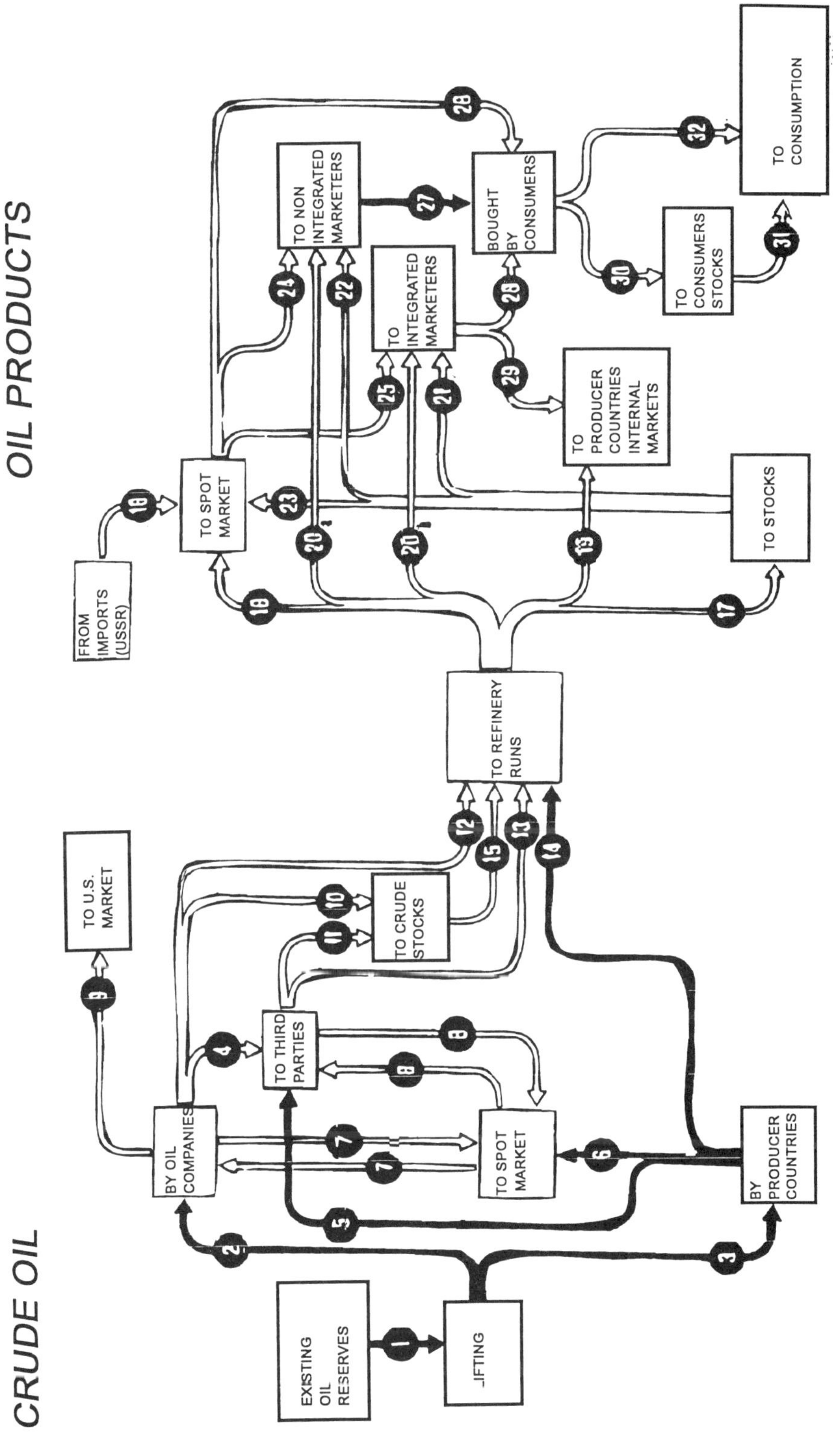

Figure 3.1 Oil production flow-lines

Figure 3.2 World flows of crude oil in 1980 (million barrels/day)

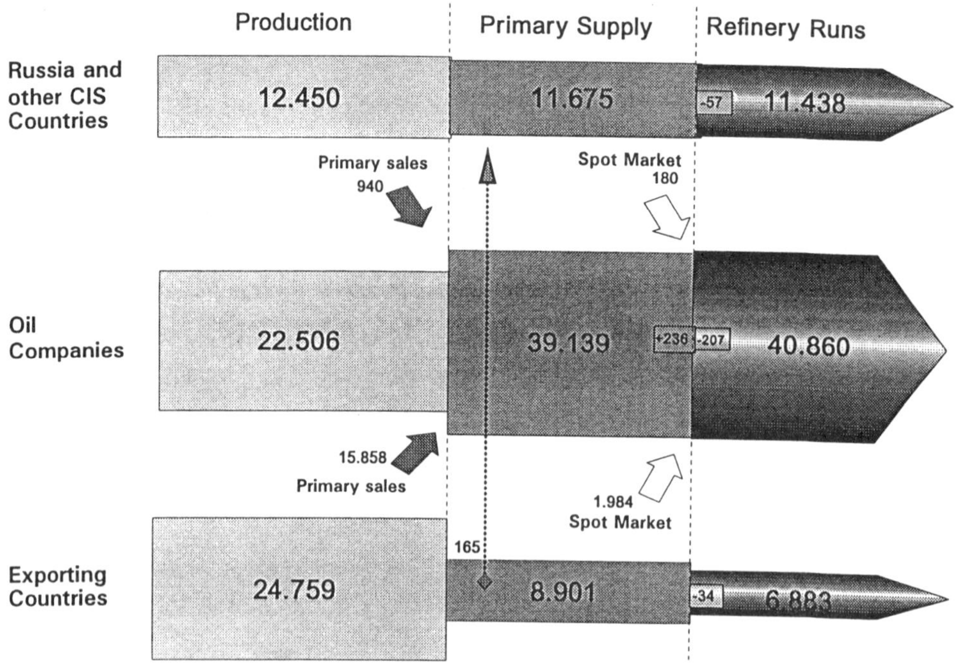

Figure 3.3 World flows of crude oil in 1983 (million barrels/day)

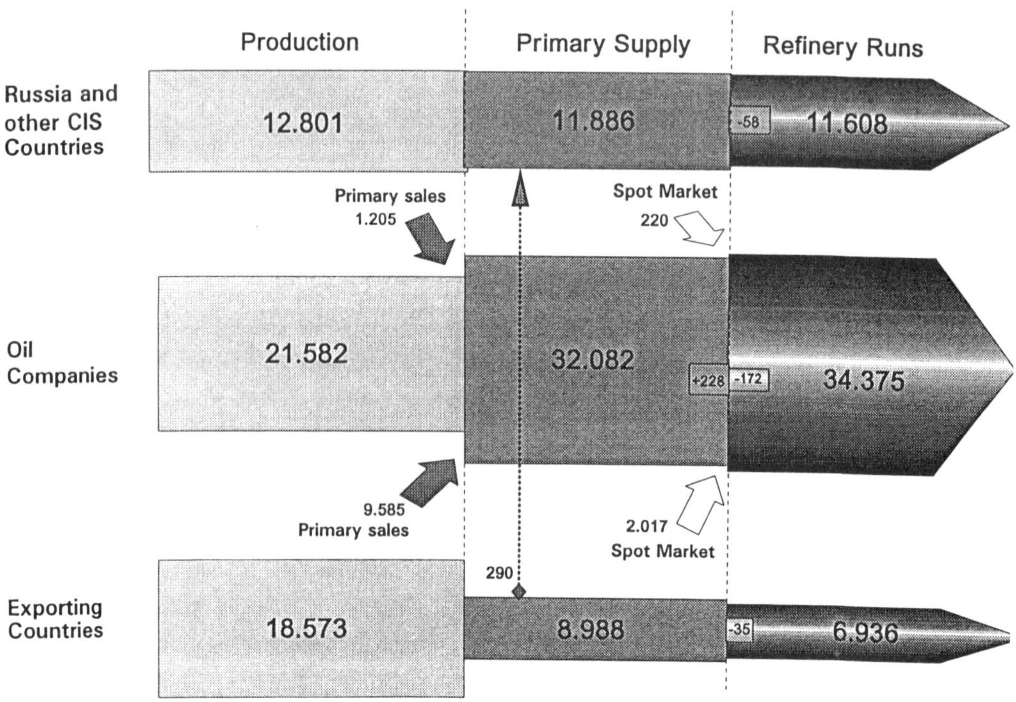

Figure 3.4 World flows of crude oil in 1986 (million barrels/day)

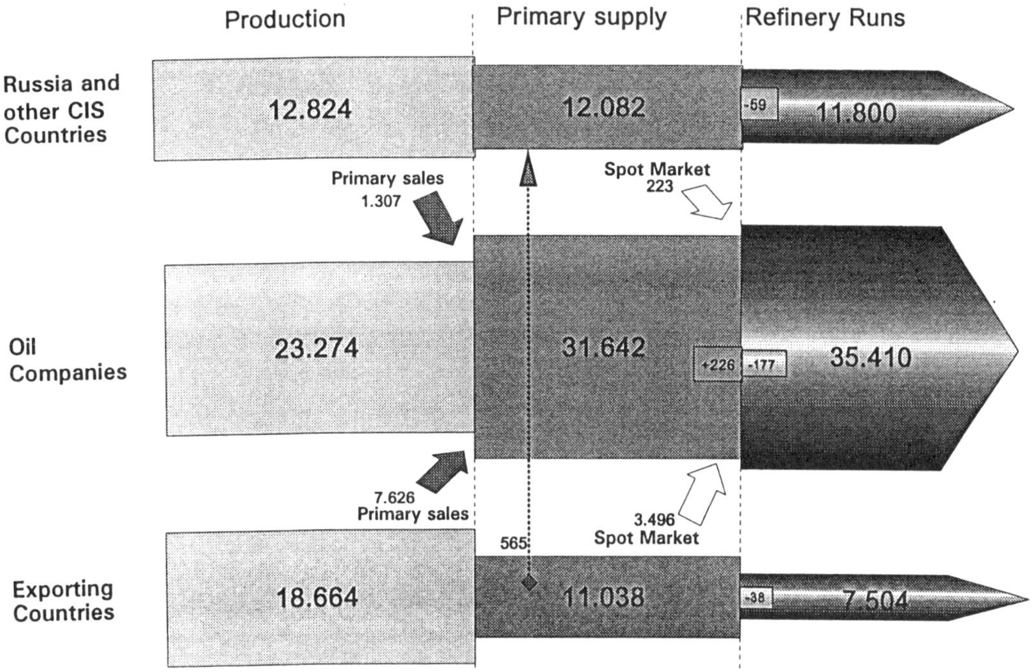

Figure 3.5 World flows of crude oil in 1992 (million barrels/day)

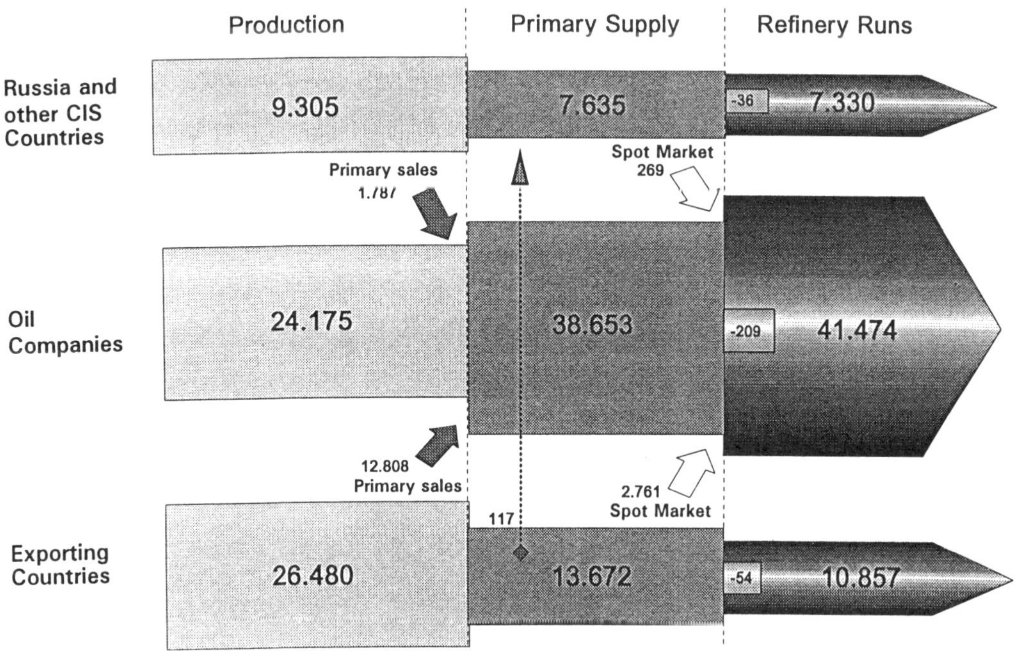

Figure 3.6 Year 1980 – World flows of crude oil (thousand barrels/day)

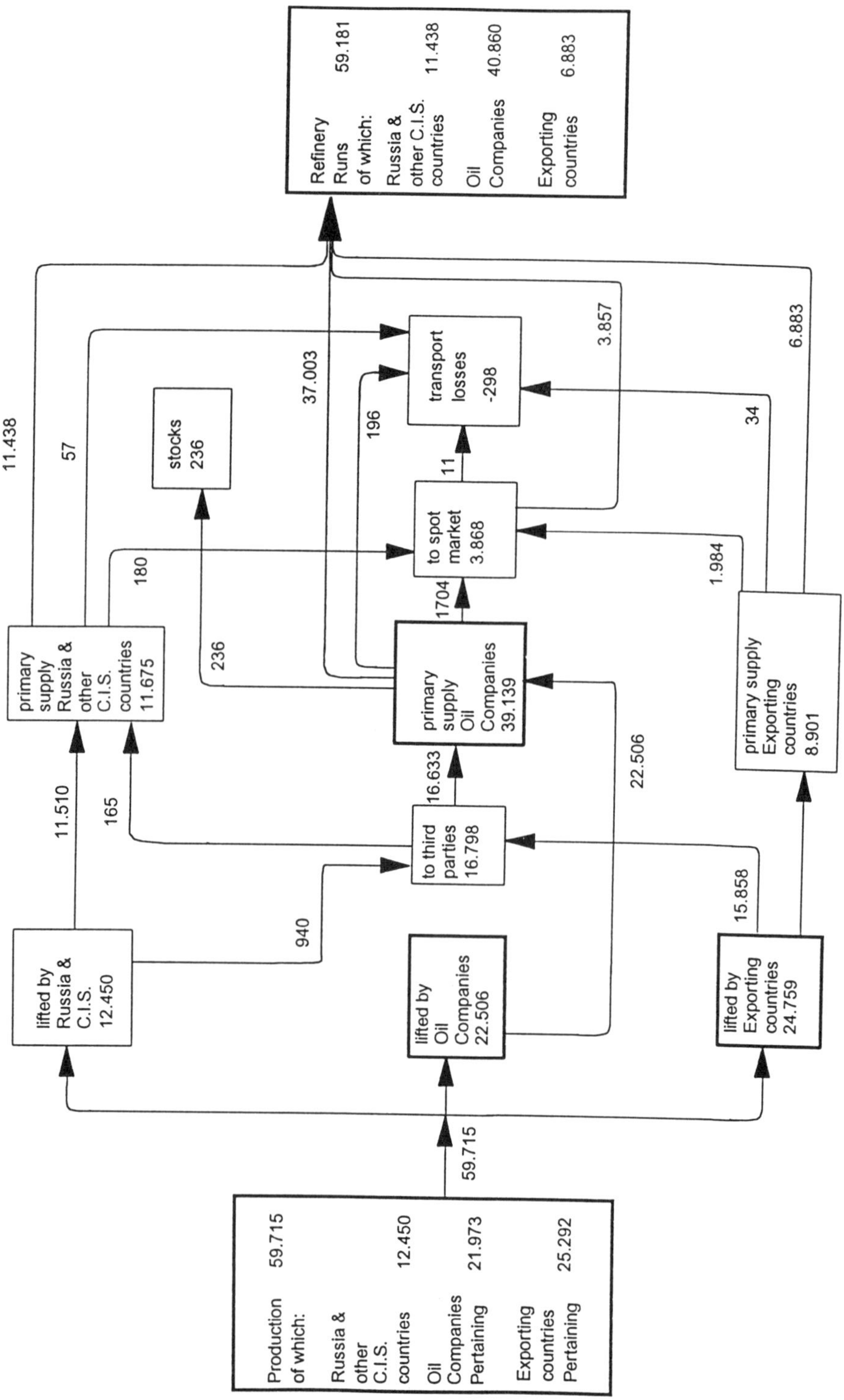

Figure 3.7 Year 1983 – World flows of crude oil (thousand barrels/day)

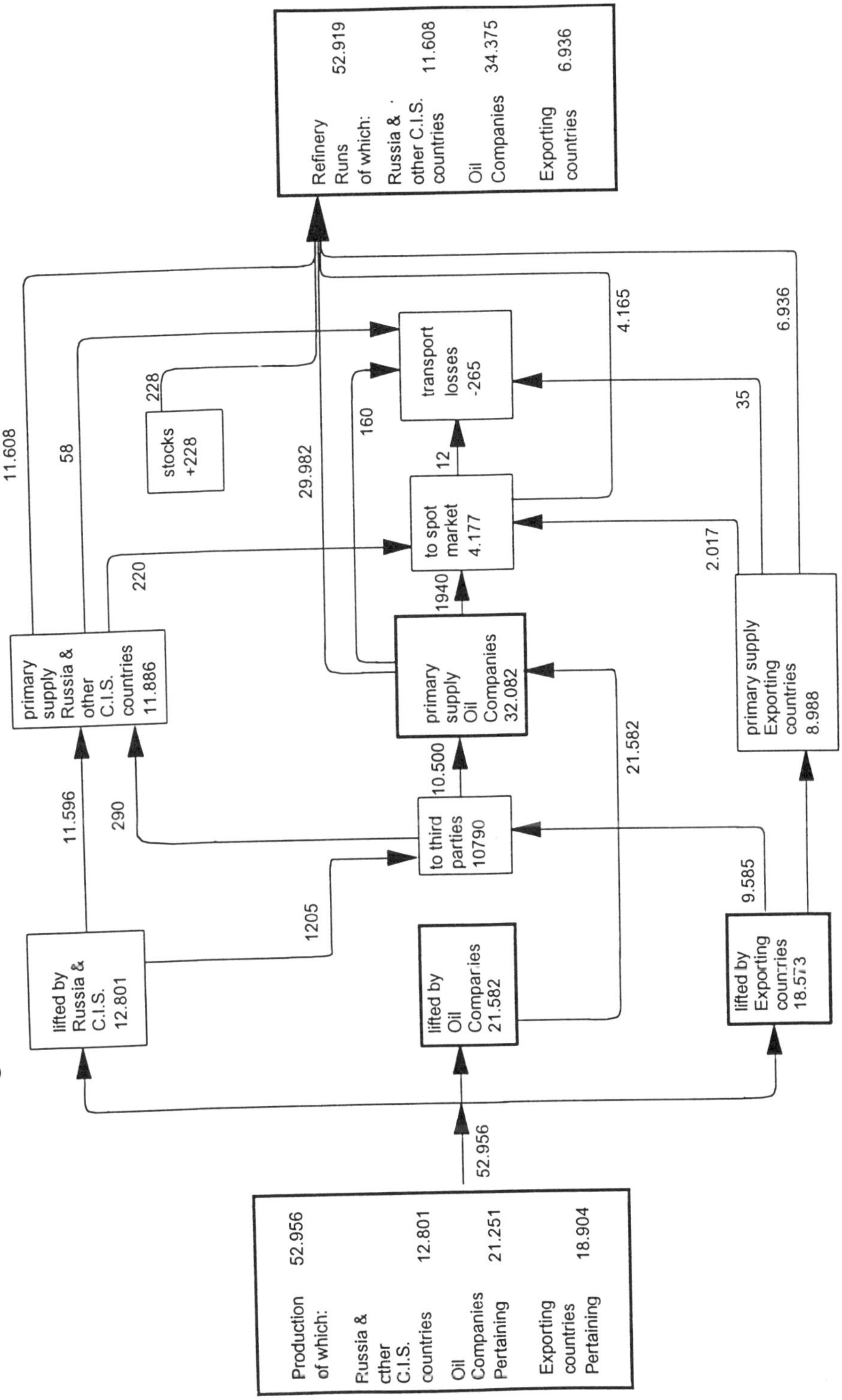

Figure 3.8 Year 1986 – World flows of crude oil (thousand barrels/day)

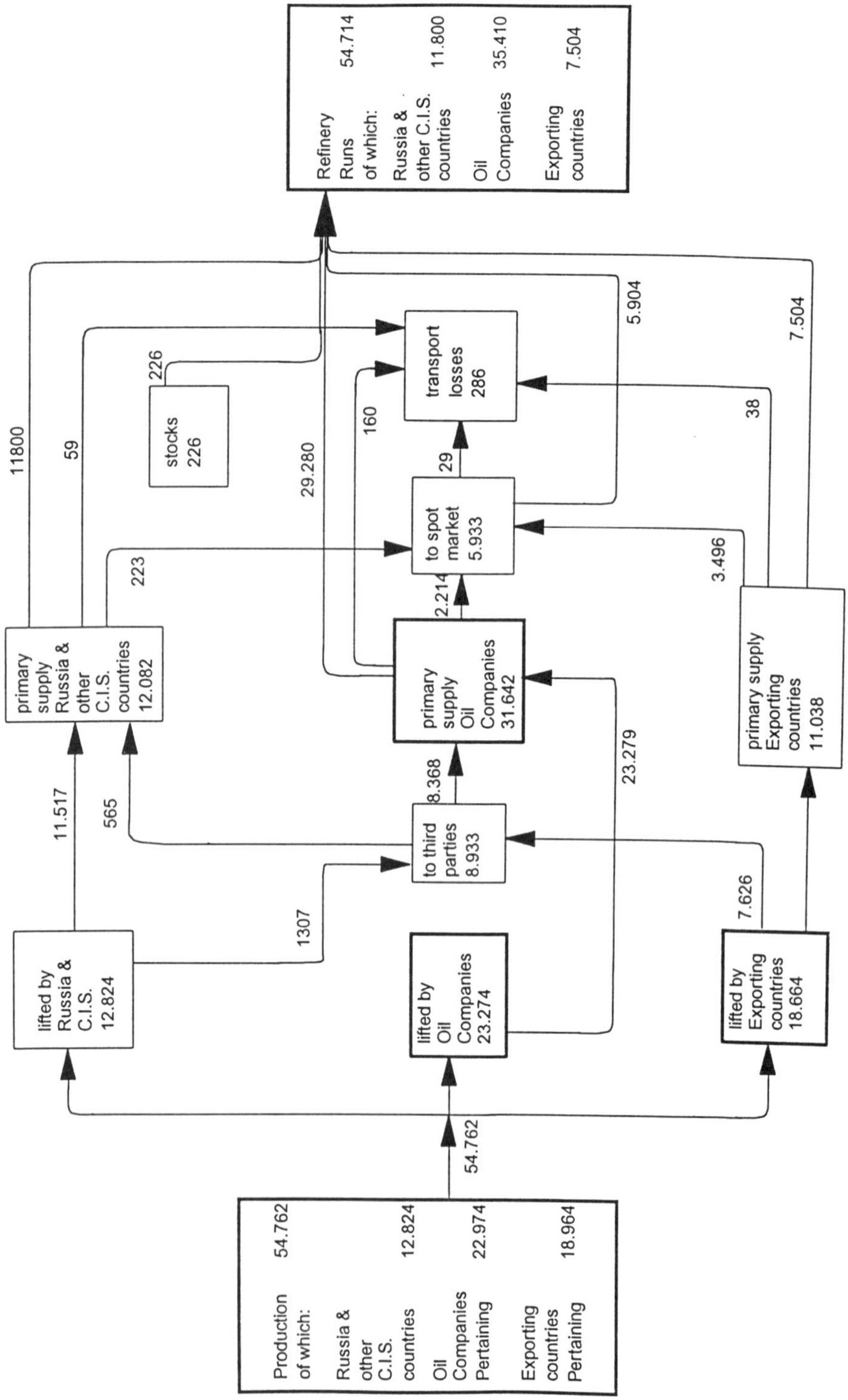

Figure 3.9 Year 1992 – World flows of crude oil (thousand barrels/day)

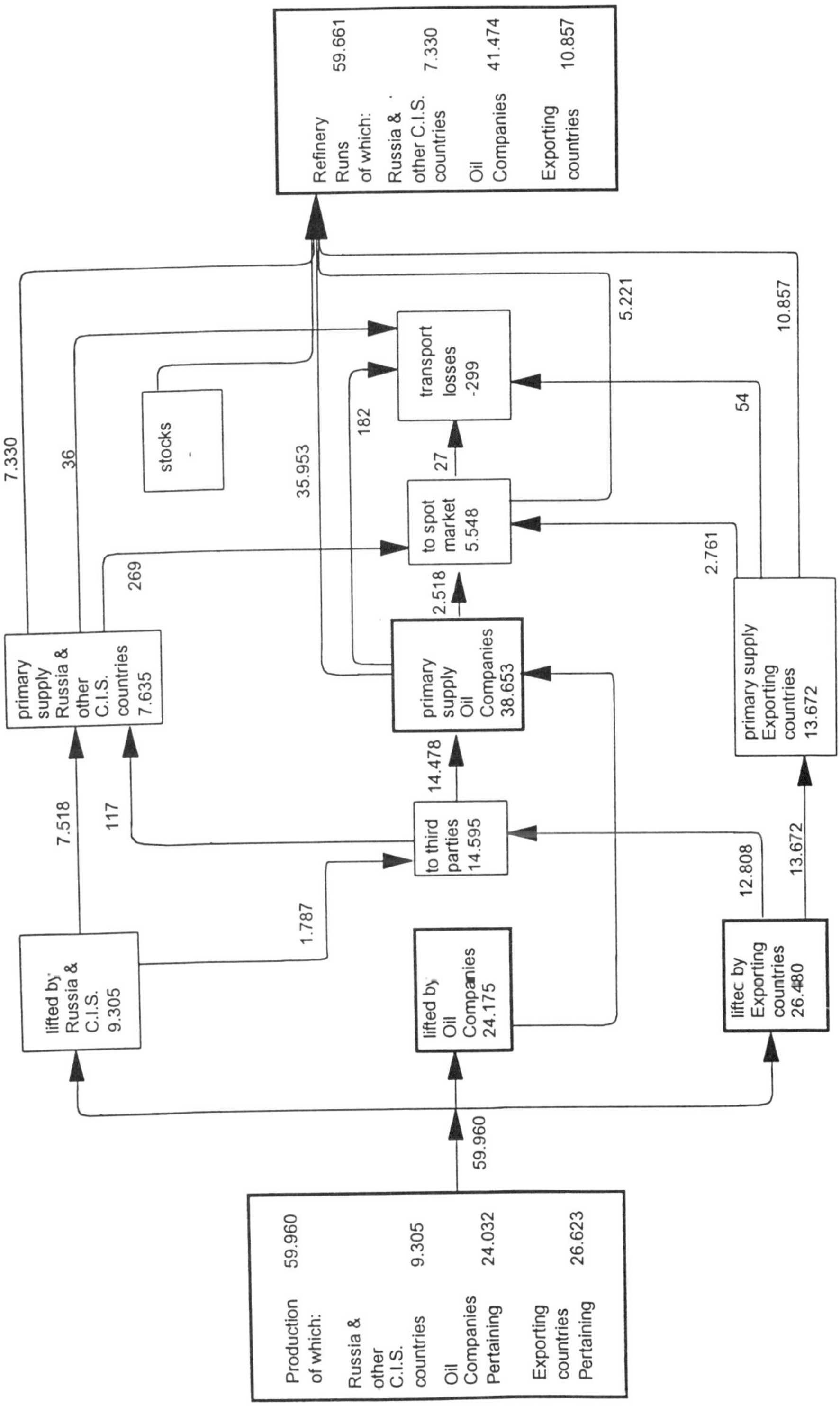

oil through national oil companies, controlled by an oil ministry, and fully integrated in the national state. Other countries of the same group: Libya, Egypt, Nigeria, etc., have given concessions to private (which means foreign) oil companies to look for, and eventually exploit, oil reserves; under the terms of concessions these companies have some right to the crude they have found.

The third group are the oil companies, large industrial organisations – private or owned by some industrial nation state – which look for crude, produce and refine it, and sell the products. They operate all over the world, in all countries which accept them.

The production figures for the exporting countries are made up of the production of their own national oil company, plus whatever oil pertains to such countries or their national oil company out of their shares in joint ventures, production sharing agreements and concession agreements with private companies.

In the same way the production figures for the oil companies are made up of whatever comes from their concessions and their shares in joint ventures and production sharing agreements in their own countries of origin or abroad.

These definitions reflect the institutional structure of the oil industry. Oil production is often a monopoly of the state company operating through its own company: alternatively, it is regulated on the basis of agreements between the government of the country – and/or its own state oil company – on the one hand and one or more foreign oil companies on the other.

These agreements often take the shape of "joint ventures" or "production sharing agreements", in which each participant is entitled to lift a quota of the oil in place roughly proportional to his share in the venture. The oil volumes which each actor is entitled to by virtue of his share in the venture are called "lifting entitlements" and the oil is called "pertaining crude". Such agreements are normally administered with flexibility, and companies may be allowed to "overlift" or to "underlift" if they need more or less oil than they have contracted for.

If at a given moment the total capacity to produce cannot be increased at short notice, or if the market will not increase its demand of that given oil, the amount "overlifted" by any private company will reduce the amount lifted by the state's company.

This difference is reported by the "spaghetti" graphs. The oil companies have overlifted in all the 4 years considered, but the volume so obtained has been decreasing, from 2.4% of the whole pertaining oil in the first to the 0.6% in the last year. This can be interpreted as a measurement of how "tight" the oil market was; that is, how supply was kept low with respect to demand.

A look at the missile-like graphs will tell the reader that, in all the 4 years considered, the exporting countries produced more than anyone else: 24.8 million barrels per day in 1980 out of a total of 59.7 mb/d. Their production, however, went down in 1983 to 18.6 mb/d out of a total of 53.0 mb/d, and remained at the same level in 1986, to jump again in 1992 to 26.5 mb/d out of a total of 60.0 mb/d. The second-largest producers are the oil companies, whose production is much more stable: 22.5 mb/d in 1980, 21.6 mb/d in 1983, 23.3 mb/d in 1986 and 24.2 in 1992. The third producer – Russia and other CIS countries – has kept its level at 12 plus mb/d for more than 10 years to crash down to 9.3 mb/d in 1992.

The production figures are, however, only the starting point.

One of the three actors in the graphs, the oil companies, need much more oil than they can produce. They have a large refinery capacity, and large shares of the consumers' product markets. They have to buy oil, and they buy a lot from the exporting countries and some from Russia and CIS. In this way, through what we call "primary sales", about 20% of the crude oil produced in the world changes hands. The primary sales are by far the single most relevant phenomenon in the oil industry's present set-up, and the reasons for this will be carefully explained in this part of the book.

In this way an average of 12.9 mb/d moved every year from the exporters to the oil companies (see Tables 3.1 and 3.2) through commercial agreements, normally "contracts" which last for some time – a year or more – and constitute the normal way to exchange large quantities of oil in a regular fashion. Primary sales were very high (28% of production) in the first of the 4 years we list, and declined very sharply in the intermediate years; finally in 1992 they climbed back to 24% of the total in 1992, as market policy of the producers eased up.

The exporting countries sold with their primary sales a large percentage of their production: 64.1% in 1980, declining to 40% in 1983 and going up to 48% in 1986 (these figures include the small volumes sold to CIS countries). Much less important are primary sales for the CIS countries, which sold 7.5% of their production in 1980, increasing as a result of an export drive to 9.4% and 10.2%, and reaching in 1992 a very high level, 19%, because of the sharp fall in their production.

The oil companies bought mainly from the exporting countries, from 40% in 1980 to 30% of their supply; and only marginally from CIS, which supplied less than 5% in any of the 4 years.

Primary sales decreased in importance throughout the period up to 1986, due to the "tight supply" policy of the exporting countries; the relaxation of this policy brought them up again almost to the level of the starting year. The following chapters will discuss in detail such changes, which have obviously been of the highest importance.

After primary sales we find in our spaghetti graphs the spot market and the stocks. Both need some explanation.

In the spot market (so-called because shiploads are bought and sold "on the spot" while travelling), the quantities of crude oil are sold and bought which for whatever reasons are not included in the "duration" contracts.

The main feature of the spot market is its lack of transparency. The same cargo

Table 3.1
PRIMARY SALES OF CRUDE OIL
from 1980 to 1992

	1980		*1983*		*1986*		*1992*	
	10^3 b/d	*%*	*10^3 b/d*	*%*	*10^3 b/d*	*%*	*10^3 b/d*	*%*
TOTAL	16,798	100.0	10,790	100.0	8,933	100.0	14,595	100.0
Primary sales of exporting countries of which:	15,858	94.4	9,585	88.8	7,626	85.4	12,808	87.8
– To oil companies	15,693		9,295		7,061		12,691	
– To CIS	163		290		565		117	
Primary sales of CIS to oil companies	940	5.6	1,205	11.2	1,307	14.6	1,787	12.2
Percentage of primary sales of exporting countries on their production	64.1		51.6		40.9		48.3	
Percentage of primary sales of CIS on their production	7.5		9.4		10.2		19.2	
Percentage total primary sales to total production	28.1		20.4		16.3		24.3	

is sold and bought many times, especially if it starts its life in the hands of a broker, who has to sell it as such, very often to another broker. It may take two, three or thirty transactions before a cargo finds a proper destination, and goes on to the next step of the oil line.

This creates a "grey area" on the market: it is impossible to say how much of the oil sold by the producers to the oil companies, and exchanged among the latter, goes through transactions of the spot type.

For this reason we have put the figures in such a way that the spot market does not change the general picture. That is, we have taken the assumption that companies exchange crude among themselves, and their refinery runs are more or less equal to their primary supply. This assumption is, after all, borne out by

Table 3.2
PRIMARY SUPPLY OF THE OIL COMPANIES

	1980		1983		1986		1992	
	10^3 b/d	%	10^3 b/d	%	10^3 b/d	%	10^3 b/d	%
Primary supply of oil companies	39,139	100.0	32,082	100.0	31,642	100.0	38,653	100.0
of which: Production of Oil Companies	22,506	57.5	21,582	67.3	23,274	73.6	24,175	62.5
Bought	16,633	42.5	10,500	32.7	8,368	26.4	14,478	37.5
of which: Bought from exporting countries	15,693	40.0	9,295	29.0	7,061	22.3	12,691	32.8
Bought from CIS	940	2.5	1,205	3.8	1,307	4.1	1,787	4.7
Percentage primary supply of oil companies on world production	65.5		60.6		57.8		64.5	

our estimates of refinery runs.

The spot market had originally been used to correct some temporary imbalance, or some planning mistake; but its importance did increase in time under pressures which made it in a sort of second-level market.

First the fiscal regulations introduced in the mid-1970s on North Sea crude did induce the producers of that crude to sell it "on the market", to provide a price usable for fiscal transactions. The North Sea crude was the first to be sold and resold, cargo by cargo, often to be bought back by the original producer. This paved the way for that crude to become a sort of market indicator, which would move at the slightest change in the supply–demand balance.

North Sea companies were not the only sellers and buyers in the spot market. In the time of short supply, i.e. from 1975 up to 1985–86, spot sales were used by oil exporters to test the market and its prices. Spot market became the marginal supplier, and its price the "market" price. These sales, and the North Sea ones, in time became quite large and important, but our figures do not really show it: quantities sold on the spot, excluding all duplications, went up in our figures from 6.5% of world production in 1980 to 7.9% in 1983, to a peak of 10.8% in 1988 only to go down in 1992 (9.2%). By its very nature the spot market is very volatile.

The quotations which developed in time in the spot market moved up and down according to the supply–demand balance; that is, the balance between the crude being produced and offered for sale by exporting countries compared with the demand expressed by the oil companies. In the 1970s and early 1980s any overlifting by the countries did end up on the spot market to test the market's reaction to higher prices, and to influence expectations. The influence of expectations on the spot price eased the transformation from a spot to a futures market.

This means that between 1980 and 1986 a company could obtain the crude it needed by paying either the technical and fiscal cost to produce its pertaining crude; or the OSP for the crude bought with contracts; or the spot price; or, finally, the cost of destocked oil.

Prices float day by day, and cargo by cargo, following in real time the changes in demand and supply. This increases the risk for any company relying on the spot market for its crude supplies, and encourages buyers and the sellers t resort to hedging, a practice common in security and commodity markets.

This has attracted brokers – that is, traders who buy to resell – and financial operators looking for speculative profits.

In a relatively short time the spot market came to include future markets and a forward – both a "paper" market and a "physical" one – market for crude oil and oil products, so that quotations have to be read together with the expected date of delivery.

In a short time the volume of financial resources invested in oil trading became enormous, and the volume of crude traded far exceeded daily world production. Phrases like "Wall Street oil field" and "paper barrels" became fashionable.

The oil market entered into a Stock Exchange – some say a casino – atmosphere, which added an uncertainty factor to the planning of companies or to the long-term policy-making of the exporting countries.

The spot market is not the only "grey area" of our representation of the market: stocks are another element not easily integrated in flows. Normally, stocks are kept to smooth operations along the logistic line, as flywheels to keep the system working during technical or logistic stoppages. They would be kept at the minimum level sufficient to prevent refineries from ceasing to operate after, say, a week of rough sea, or pipelines from running dry if one well blew up.

In a volatile oil market, however, stocks became another element of complexity. If a company expects the price to rise between now and, let us say, the end of the year, it will presumably keep whatever stock it has, and also buy some extra, which, by the way, produces the very rise in prices that it has been expecting. In this way the company will be able to use the oil in stock either to sell it at a profit, or to supply its refineries with crude cheaper than the current price.

On the contrary, if a company expects prices to go down, it will sell whatever

Table 3.3
CONTROL SHARE OF THE OIL BUSINESS BY THE MAIN ACTORS (percentages)

	Russia and CIS	*Oil companies*	*Producing countries*	*Total*
Production				
1980	20.8	37.7	41.5	100.0
1983	24.2	40.8	35.0	100.0
1986	23.4	42.5	34.1	100.0
1992	15.5	40.3	44.2	100.0
Primary supply				
1980	19.6	65.5	14.9	100.0
1983	22.5	60.6	16.9	100.0
1986	22.1	57.8	20.1	100.0
1992	12.7	64.5	22.8	100.0
Refinery runs				
1980	19.3	69.1	11.6	100.0
1983	21.9	65.0	13.1	100.0
1986	21.6	64.7	13.7	100.0
1992	12.3	69.5	18.2	100.0

stock it has in excess of strict operational needs. Otherwise it would fill its tank farm with oil more costly than the current price, which could not be processed, or sold, without loss. It will also refrain from buying stocks, which will be cheaper later. As the total volume of stocks kept world-wide is very large, expectations of a price drop would create a destocking movement large enough to create an additional oil supply. The opposite, an artificial shortage, could also happen. The market becomes easy to manipulate.

Of course, stock management is not easy: it has to consider, among other elements, the cost of keeping stocks, which moves with the rate of interest, and became very high after the price increases of the 1970s. Companies may sell stocks for all sorts of reasons: because they need some profit-taking, or because they have to increase their cash flow for budgetary or financial short-term reasons, etc. A particularly thorny problem is the change in the value in stocks written in balance sheets due to changes in the oil price. The 1986 fall in oil prices sent shivers down the spines of accountants and controllers, as well as chairmen, faced with the need to devalue the stocks at the end of the year, thus creating huge patrimonial losses.

Stock build-up can be the result of overlifting. A company facing low demand for its crude may be bound by lifting commitments it cannot or does not wish to

Table 3.4
WORLD FLOWS OF CRUDE OIL

	Production	*Primary sales*	*Primary supply*	*Spot market*	*Stocke*	*Transport losses*	*Refinery runs*
Total 1980	59.715	–16.798	59.715	–3.868	–236	–298	59.181
Russia and other CIS countries	12.450	–940	11.675	–180		–57	11.438
Oil companies	22.506		39.139	–1.704	–236	–207	40.860
Exporting countries	24.759	–15.858	8.901	–1.984		–34	6.883
Total 1983	52.956	–10.790	52.956	–4.177	+228	–265	52.919
Russia and other CIS countries	12.801	–1.205	11.886	–220		–58	11.608
Oil companies	21.582		32.082	–1.940	+228	–172	34.375
Exporting countries	18.573	–9.585	8.988	–2.017		–35	6.936
Total 1986	54.762	–8.933	54.762	–5.933	+226	–286	54.714
Russia and other CIS countries	12.824	–1.307	12.082	–223		–59	11.800
Oil companies	23.274		31.642	–2.214	+226	–189	35.410
Exporting countries	18.664	–7.626	11.038	–3.496		–38	7.504
Total 1992	59.960	–14.595	59.960	–5.548		–299	59.661
Russia and other CIS countries	9.305	–1.787	7.635	–269		–36	7.330
Oil companies	24.175		38.653	–2.518		–209	41.474
Exporting countries	26.480	–12.808	13.672	–2.761		–54	10.857

break, and will have no choice but to stock its extra production, perhaps leaving such stocks in floating storage if its tank farms are full.

Moreover, crude stocks are of public interest, and governments and the International Energy Agency may frown upon juggling with something they consider essential to prevent a supply crisis.

It seems clear that stock management is basically related to expectations, which can be manipulated up to a point; and that the original function of stocks, to reduce uncertainty, has been reversed. Today the movements of stocks anticipate and amplify market movements, and increase the overall uncertainty and complexity of the oil market.

This somewhat tiring explanation of spot market and oil stocks has led us to the end of the spaghetti junction: to the refinery runs, which have been larger in 1983 and 1986 than in 1980, and were in 1992 at the same level of the first year.

We can now perceive the full meaning of the present oil set-up by considering the relative "shares of control" of the three main actors. The figures in Tables 3.3 and 3.4 show that the oil exporting countries, which in 1992 controlled 44% of the production, reduce their share of control to 23% at the level of primary supply (that is, after primary sales) and reduce it further to 18% at the level of refinery runs.

The latter figure is, however, much higher than in the recent past, due to the effort of the exporting countries to increase their share of refining, starting from the very low level of 12%.

The oil companies' share of control moves in exactly the opposite way: their 40% control of the production becomes 65% after the primary sales, and 70% in the refineries, the latter a very high level due mainly to the plunge taken by CIS countries, which had made an effort to increase their participation to production, which was increased from 20.8% to 23.4% in 6 years, and then let it drop in 1992 much below the starting point; the same has happened in other two phases of the oil business.

CHAPTER 4

THE "GOOD OLD DAYS"

The need to "manage" the oil industry

When it starts producing from a new field an oil company has already paid for the exploration and invested in the development of the field, and is therefore heavily out of pocket. It now has to recover this investment and to make a profit out of it.

Oil companies invest for profit, and as a dollar today is worth more than a dollar tomorrow, they want to get the money back as soon as possible. To do this they will adopt – if the market allows – the quickest depletion rate compatible with the need to recover the maximum amount of oil from the field and to lift as much oil as possible using the natural energy (i.e. pressure) of the field. When this energy is exhausted, costly artificial pressure will have to be created and maintained. Before that moment, once you have put the field into production, the out-of-pocket cost of producing the oil is usually very small.

If oil demand is high enough and/or the oil field is too small to influence the overall supply–demand balance, the field will be depleted at the maximum efficient rate of production (MER), the quickest rate compatible with the highest recovery.

However, if the field is very large and/or demand is low, production at MER would flood the market and depress prices: a lower rate of depletion will have to be adopted and the field will last longer; other things being equal, its cash flow will be smaller. For these reasons, oil companies often cannot exploit the larger fields as fast as they would like. These fields can only be depleted slowly, and their potential capacity – and even their effectively available capacity – are higher than production. For this reason, and because of the low out-of-pocket production costs, the oil industry is characterised by a low short-term marginal cost: that is, the cost of an additional barrel from existing production capacity is very low.

Existing fields competing with other fields would therefore continue to produce until the price falls below their direct costs, i.e. lower than direct out-of-pocket expenses. Going to extremes, it could be said that there is no price, however low, which can force an existing production to cease; and that supply is adjusted to demand not through price changes but by accelerating or slowing down the development of new capacity. Low prices discourage exploration, and cause the development of new fields to be postponed, so that supply does not keep pace with demand, and prices go up. However, this mechanism is quite clumsy. Cheap fields may be developed even at low prices; supply might be reduced too gradually

M. Colitti and C. Simeoni, Perspectives of Oil and Gas: The Road to Interdependence, 65–72.

to have an immediate influence on the short term. In other words, the oil industry has to be "managed", and in fact it always has been. It cannot simply be left to the market as any other industry might, not just because of its size or the importance of its product, but because of its economic structure and the peculiar shape of its cost curve, which would in any case create a condition of oversupply, and therefore of very low prices, followed by supply crises. To prevent this, the adjustments must be managed. It would indeed be difficult to express the question better than Paul Frankel, who in 1973, wrote: "My own assessment of oil problems is based on the characteristic features of a highly capital-intensive industry, an assessment spelled out in my *'Essential of Petroleum'*, first published in 1946, and in a number of later publications of mine. Throughout I have traced the all-pervading repercussions of the extreme relation of fixed and variable costs which in inverse ways determine the operators' actions and reactions: on the one hand the rapidly falling cost curve – which moving towards the tail end of the available capacity the marginal costs are low compared with those of the baseload – encourages intensive competitive behaviours; on the other, and for that selfsame reason, there is a deepseated tendency to avoid, or at least to mitigate, the results of this objective situation not only by horizontal concentration and vertical integration, but also by co-operative endeavours leading to varying degrees of "understanding" among competitors – some of them engendered by governmental authorities. I thus come to the result that the oil industry (mainly due to its built-in anarchic features) is not self-adjusting and therefore tends to generate adjusting measures."[1]

The best historical example of such an adjustment is the discovery in the 1930s and 1940s of the giant fields of the Middle East. They were so large that at MER they would have exceeded the market they could have reasonably hoped to supply for any length of time. The oil there being so abundant and cheap, the limitation (sometimes called "proration") did not reduce the companies' profits, which were in fact quite high. Even with strict limitation, the Middle East production was large enough to create a market in which oil was sold at a price that appropriated the oil rent to the cheapest producer; i.e. at prices more or less equal to the marginal supply cost.

In this way, the Middle East oil would have found its outlet in the largest consuming market, the USA, by successfully competing with the more costly local oil. However, it did not happen. Local producers resisted the prospect of being displaced, and they had enough political power to isolate the US market from the rest of the world. In the early 1960s the US crude was kept at a price that covered the high cost of local production, and the Middle East oil was relegated to the market outside America, mainly Europe and later Japan. To quote P.

---

[1] See Paul Frankel, *Common Carrier of Common Sense*, Mabro and Skeet, editors, Oxford Institute for Energy Studies, p. 214.

Frankel again (p. 215): "Obviously the full appraisal of the volume potential of the Middle East reserves, and of a cost level much lower than anything previously envisaged, led to a downward turn of what could previously be considered to be the prevailing level of prices. It was the constitution of the oil industry, and its political environment, which inhibited the fuller and prompter spread of these lower costs."

The giant fields were in fact "prorated" according to European demand. This was the first case, which was repeated later at OPEC's decision, of inverting depletion: the high-cost, declining reserves of the USA were exploited in preference to the low-cost, long-term reserves of the Middle East.

Although the European market grew quickly enough to satisfy the companies' profit expectations, the division of the world market remained a sore point, and the multinationals never really accepted it. During the early 1970s they tried again to enter the USA, and their rejection was perhaps one of the reasons for a relatively mild attitude adopted by the multinationals in the negotiations of the early 1970s with OPEC countries.

Management by integration and interlocking concessions

The history of oil starts at the end of the last century in the USA, where a multitude of entrepreneurs started producing oil and selling it to transporters and refiners. The latter, who started very early to grow in size, used to "post" – that is, to nail on their doors – the prices they would pay for the crude. The original price was therefore called "posted price", a name which lasted a long time after that structure of the industry had disappeared.

By the 1930s the production and refining of oil and the distribution of oil products had fallen almost totally into the hands of a few large integrated oil companies variously known as the "seven sisters", "the majors", or "the multinationals". They used to run the oil fields on the basis of concessions granted by the political authority of the land. Outside their countries of origin, this was very often a colonial or semi-colonial government.

The multinationals, the "seven sisters" as Enrico Mattei would call them, operated all over the world, and participated with varying shares in practically every oil area in the world. Saudi Arabia, for example, was the preserve of four large US Companies, Exxon, Mobil, Chevron and Texaco; the Iranian reserves, after the aborted attempt at nationalisation by Mossadeq, were exploited by an international consortium led by BP (originally called Anglo Persian), etc. (see Figure 4.1).

The same formula was repeated whenever it was deemed practical. There were at least two important reasons for this. A joint venture, i.e. an association in which investments and production are divided according to each participant's

Figure 4.1 International consortia in the Middle East

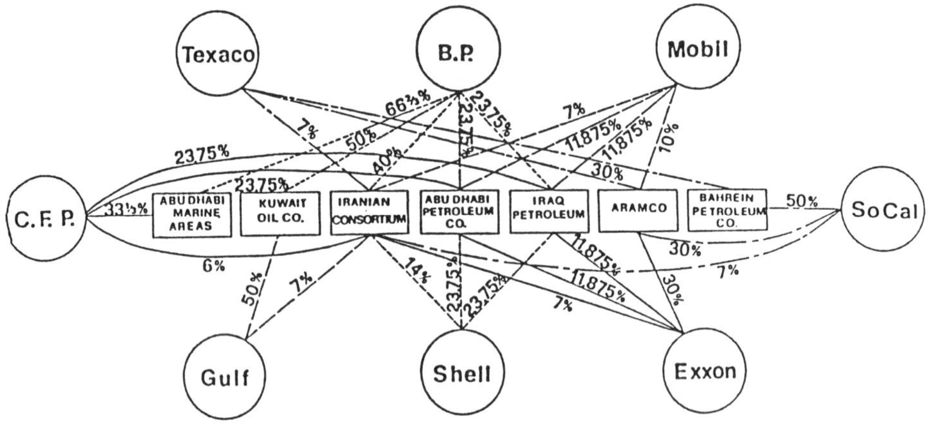

Source: US Senate, Hearings before the Subcommittee on Multinational Corporations of the Committee on Foreign Relations, multinational petroleum companies and Foreign Policy, Part V, p. 290

share, reduces the overall mining and political risk of its members, and subdivides their capital commitment. In this way each company can participate in many projects, and increase the probability of success without stretching its financial resources. "Interlocking concessions", as they were called, enabled the companies collectively to control oil production outside the USA, by agreeing on the level of production of the various consortia in which they participated. Basically, it was an expression of their dominant position, as well as a very strong instrument to maintain it.

Table 4.1 lists the participants in the four most important consortia outside the USA and the main international companies' shares in them. The four Consortia alone produced in 1970 12.2 mb/d, 43% of the total oil produced outside the USA and USSR; which was enough to give the majors control of supply and prices world wide.

According to the Committee on Foreign Relations of the US Senate[2], which reported data taken from the *OPEC Statistical Bulletin*, the seven majors controlled in 1972 70% of the total amount of oil produced in the world, excluding Eastern Europe and China. The companies produced at the same date 39.1% of the US oil and 77% of the oil from the Middle East. At that time "posted

---

[2] National Corporations and United States Foreign Policy: Hearings before the Subcommittee on Multinational Corporations of the Committee on Foreign Relations – USA Senate – 93th Congress – Second Session on Multinational Petroleum Companies and Foreign Policy, 30 January 1974

Table 4.1
SHARES OF THE MAJOR INTERNATIONAL OIL COMPANIES IN THE PRODUCTION CONSORTIA IN THE MAIN OIL AREAS OUTSIDE THE USA

	Country/company			
	Saudi Arabia: Arabian American Oil Company (ARAMCO)[1]	*Kuwait: Kuwait Oil Company (KOC)*[2]	*Iran: Iranian Consortium*[3]	*Iraq: Iraq Petroleum Company*[4]
Chevron Socal	30	–	7	–
Exxon	30	–	7	11,875
Texaco	30	–	7	–
Mobil	10	–	7	11,875
B.P.[5]	–	50	40	23,750
Gulf	–	50	7	–
Shell	–	–	14	23,750
CFP	–	–	6	23,750
Partex[6]	–	–	–	5,000
Others	–	–	5	–
Totals	100	100	100	100,000

[1]In 1973 the Saudi Arabian government acquired a participation of 25% in ARAMCO (Arabian American Oil Company); in 1974 the government increased the participation to 60% and in 1980 to 100%.

[2]In 1933 Gulf Oil and the Anglo-Persian constituted the joint venture called KOC and concluded with Shell a long-term supply agreement. In 1975 the Kuwait government took 100% control of KOC.

[3]In 1974 the Iranian government took total control of the Iranian consortium.

[4]The Iraq Petroleum Company was created in 1928. The nationalization of IPC was started in 1972 and completed in 1975.

[5]Anglo Persian from 1908; Anglo Iranian from 1935; BP from 1974.

[6]Gulbenkian's company.

Algeria in 1971 acquired 51% of the equity crude of French companies CFP Total and ELF.

Libya in 1971 nationalized BP's equity crude; in 1973 it acquired half of AGIP's and nationalized Hunt's, OXY's and Oasis Group's equity crude.

Venezuela in 1976 nationalised its oil industry.

Nigeria in 1976 nationalised 55% of its oil industry.

price" had come to mean a price decided by the oil majors and used for tax purposes in agreement with the producing countries' Governments; and also to calculate the discounts that the majors would offer each other.

The second key element in achieving the same goal was integration.

Under the majors' control oil flowed only along their lines of integration, i.e. through their transport systems, into their refineries and distribution networks. Very little crude was sold as such: strictly speaking the oil market did not exist at all or, rather, it was confined to the members of a very exclusive club. Some crude oil did change hands before being turned into products to be sold to the consumers, but only among major companies.

The majors did not meet much competition in the crude production phase, which required long negotiations with foreign entities (and the necessary political support from home) as well as large resources to invest in very risky operations. Competition was of course potentially much higher in refining, when all you had to do was to build a refining plant; while in the final phase, distribution and sales, competition was limited – but not cancelled – by the time and money required to build up a road-side distribution network. The majors had a strong interest in concentrating their profits in the production phase, in which they were dominant, and to allow the other phases only enough to break even.

The downstream phases were seen simply as a way to dispose of the crude, and not as profit makers: the profits were made upstream, and the other phases made very little money, or even lost money, which was sometimes useful for fiscal purposes.

In due time, however, the club members developed in different ways. Some found large fields, and became "long", i.e. had more crude than downstream markets, and had to sell crude oil. Others became tendentially "short", i.e. developed market outlets more than production, and had to buy crude. Very few of them remained perfectly balanced, i.e. produced exactly the amount of crude they needed to feed their refineries and their marketing systems.

For some time after the Second World War, for example, Shell was short, and bought oil on a long-term basis from Gulf Oil, which was long; BP was long, while Esso was the most balanced of all.

Long-term oil contracts stipulating fixed quantities and prices (or, more frequently, based on net-back pricing) were taken as a basis for exchanges among the majors. Their duration allowed for good planning and building logistic facilities with a good knowledge of the quantities that would flow through them, to ensure that the utilisation of capacity would be high, and the cost per barrel correspondingly low. The oil bought carried little, if any, of the production profit, but at least its supply was reliable and regular.

Of course the large companies were tendentially not so accommodating when they dealt with newcomers struggling for a share of the market. The latter were not members of the club, whose main function was precisely to prevent such newcomers from developing. During the oil glut of the late 1950s and early 1960s, some of the "long" majors needed somebody to take the surplus crude that could not go through their integrated network. A "third party market" developed to supply crude to new operators like AGIP which had built a strong position downstream in Italy but was short of crude. Quotations and general terms became somewhat less rigid and throughout the 1960s discounts on crude were the rule rather than the exception. A spot market started to develop, but integration still prevented the creation of a real crude market. Although the system came under serious strain due to additional supplies in the early 1960s, it is interesting to note that in the early 1970s it was basically short, possibly due to this effect of integration rather than to lack of continuity in the exploration and discovery process.

Why the system came to be terminated

The problem was not economic or managerial, but political. The concession system was rapidly becoming untenable: its connection with colonialism denied any role in the running of the oil industry to the very countries in which the oil reserves were located. The old system came under a two-pronged attack: there were the tensions within the West, and between West and East.

The majors had in fact been embarrassed, for example, by the private American operators such as Occidental Petroleum (OXY), owned by the maveric entrepreneur-cum-financier Armand Hammer, which in the late 1960s found crude in Libya. The Italian state-owned company, ENI, was clamouring outside the club without letting anybody know whether it wanted to enter it (which no member would ever allow) or to change the rules of the game, and supersede the club altogether (which nobody thought ENI could do). However small, the position of ENI had been for some time a serious matter for the multinationals. Enrico Mattei was ready to conclude agreements with exporting countries that made them partners in the exploration venture; he also publicly accused the "seven sisters" of humiliating both the smaller European countries which, like Italy, were kept out of every international consortium, and the exporting countries, kept in a purely passive role. Mattei was also ready to buy oil from the USSR, and had actually signed a contract to that effect in 1960. ENI had answered the ensuing uproar by saying that alleged "strategic reasons" were in fact veiling the defence of the multinationals' dominant position and it had kept its ground. The USSR had become a *bona-fide* oil supplier to the West.

Finally, the oil market itself was not what the multinationals would have liked, as the practical exclusion of Middle East oil from the US market was still a matter of loss of face and, above all, of good profits. The multinationals thought they could easily stand up to competition from both American and European independents, but their inability to sell Arabian crude in the USA was an open wound; so was the oil glut that all these pressures created, in which companies without equity crude could survive and prosper by buying the available crude. Everything seemed to point to a trend towards a fall in the price of crude oil,[3] which the majors would do anything in their power to prevent.

The East-West tensions between the oil companies and the exporting countries were even stronger: OPEC had been born in 1960, as a result of pressure initially from Venezuela and then from Middle East countries.

Although the organization's early years had been spent negotiating relatively minor fiscal clauses of oil agreements, OPEC gave the producers an instrument for concerted action, and encouraged them to take more and more intransigent positions.

The old system was increasingly seen in the Middle East and in Venezuela as a remnant of the colonial regimes which had been spectacularly terminated, at least in Africa, in the previous decade. Something which was eventually called "the Third World movement" was actually much stronger than people in the rich countries thought, and the last colonial war had recently been lost by the French in Algeria. The spirit of the times clearly pointed in that direction, and the dollar devaluation of 1971 gave the movement urgency and consistency, as it was widely read as a proof the weakness of the USA, and as a sign that the producers of raw materials could not expect a steady income even in a demand boom.

The main ingredients for a blow-up were all there.

[3] The famous guess by M.A. Adelman in his book, *The World Petroleum Market* (John Hopkins University Press, 1972) of a \$1/b crude oil was derided by almost everybody, as it came out almost immediately before the first oil shock. However, the evaluation was correct: it measured where crude oil prices would have gone without the shock. The book prompted Paul Frankel's (*op. cit.* p. 213) answer in the chapter: "The oil industry and Professor Adelman".

CHAPTER 5

THE OIL REVOLUTION

Introduction

In the early 1970s, through a clever mix of negotiation, nationalisation and cuts in production, some oil-producing countries acting alone and together in OPEC took over a business previously controlled by the multinational oil companies, and brought about a four-fold increase in oil prices. They published an official selling price (OSP) (also called government selling price – GSP) which became the price of oil, and was brought, by negotiation or unilateral OPEC decisions, to unprecedented heights. The event was called "the oil revolution" and the two jumps of 1974 and 1979 in the price "the oil shocks" (see Tables5.1). This amounted to the largest economic dislocation since World War Two, and deeply shook the oil-consuming countries, a composite group which included a big producer–importer like the USA, large consumers with little or no production like most of Europe and Japan, and the poor importers of the Third World.

The revolution deeply changed the situation of the oil exporters, and also both of those who had engineered it, i.e. the OPEC countries, and those who profited by it, the so-called non-OPEC exporters such as Mexico, the North Sea countries, Egypt and the Soviet Union.

Twenty years have since passed, and the event is by now fading from our collective memory. Its history has been told over and over again by a number of very good books, so that we do not need here to follow the facts, but rather to extract from them whatever message they may hold for us now.[1]

The oil revolution happened in two spurts, prepared and accompanied by negotiations and political decisions which form a continuum of great complexity.

Some points are, however, quite firm. By 1973 the OPEC members were producing 56% of all the crude produced in the world; their share had been on the increase for some time, and the general consensus was that it would continue to do so (see Figure 5.1).

In 1974 the oil prices fixed by OPEC jumped from \$3.7 to \$10.7 per barrel, while OPEC production went down by 300,000 b/d (see Tables 5.1 and 5.2 and Figures 5.2 and 5.3). The value of OPEC export tripled, from \$36.9 to \$119.8 billions (see Table 5.1 and Figure 5.4).

---

[1] I need to quote but one, the oil book to end all oil books: D. Yergin's *The Prize* (Simon & Schuster, New York, 1991).

M. Colitti and C. Simeoni, Perspectives of Oil and Gas: The Road to Interdependence, 73–90.

Figure 5.1 Crude oil market share of OPEC and other producers from 1970 to 1994

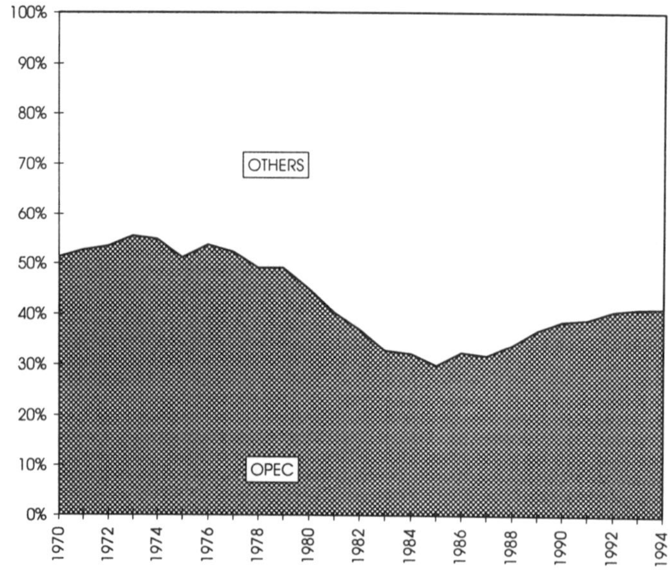

In the following year, 1975, the price remained at the high level it had reached (losing only 2 cents/barrel) while OPEC production went down by 3.5 mb/d, bringing world oil production down by 3 mb/d. Value of OPEC exports also went down by $12.5 billion to $107.3 billion.

After some years of a relative lull, the new Iranian leadership resulting from the 1978 revolution drastically cut the country's oil production, so that the total OPEC production fell from 30.9 mb/d in 1979 to 26.9 mb/d in 1980. Prices flared up again from $28.6/b to $36.0/b, bringing the value of petroleum exports of OPEC countries to an all-time peak of $279.3 billion. (The OPEC organization and the decisions settled on oil productions and prices in the 97 OPEC Conferences held from 1960 to 1994 are reported in Annex 2)

We will try here to describe the reaction of the people most affected by the revolution, which will bring us to the political side of the event. This part of the discussion will by necessity be centred on the price–demand effects, and on the adjustments after the shock rather than on the shock itself.

The oil consumers and the developed oil-consuming countries

Consumers do not use energy sources for their own sake but in order to obtain the services necessary to produce or transport goods, and to enjoy comfort in the home and while travelling. Modern life and economy would be impossible

Table 5.1
OPEC FROM 1970 TO 1993

	Crude oil price ($/b)		*Values of*		
	Nominal	*Real (base 1973)*	*Petroleum exports (in $-billion)*	*Imports CIF (in $-billion)*	*Current account balance (in $-billion)*
1970	2.6	1.9	14.5	10.1	3.2
1971	3.0	2.3	19.7	12.1	3.7
1972	3.2	2.6	23.5	13.9	4.1
1973	3.7	3.7	36.9	20.0	6.8
1974	10.7	8.5	119.8	32.3	67.2
1975	10.5	7.9	107.3	51.2	32.2
1976	11.6	8.6	129.6	63.3	37.3
1977	12.6	8.8	141.4	84.7	21.9
1978	12.8	7.9	134.4	95.7	–2.2
1979	28.6	15.0	202.0	99.6	59.5
1980	36.0	16.4	279.3	125.0	104.4
1981	34.3	16.2	257.4	147.5	52.6
1982	33.0	15.5	201.1	146.8	3.7
1983	30.3	15.0	155.1	136.0	–14.3
1984	28.5	14.5	144.1	120.9	–1.1
1985	26.9	14.2	128.9	99.6	7.5
1986	13.0	5.9	75.0	87.0	–21.4
1987	17.8	7.3	93.3	89.5	–6.4
1988	14.0	5.3	86.4	101.7	–13.8
1989	17.2	6.8	110.2	104.5	6.8
1990	22.9	7.8	144.6	114.0	20.0
1991	18.7	6.7	129.6	131.4	–55.7
1992	18.5	6.2	136.2	142.5	–22.4
1993	16.1	5.9			

Note: from January 1987 prices are based on a basket of seven OPEC crudes.
Source: Oil & Gas Journal, OPEC Annual Statistical Bulletin.

without such services; but the fraction of their cost due to energy – quite high in the case of running a car or a steam cracker – is relatively low in the economy as a whole. The industrial countries see raw materials and energy merely as a cost item, less important than the value they add to it: this is not so with the producers–exporters – where oil was, and still is – the main (or the only) modern sector of the economy, as well as the only source of foreign currency and investment capital.

Table 5.2
WORLD CRUDE OIL PRODUCTION FROM 1970 TO 1993

Years	*World production (mb/d) (1)*	*OPEC production (mb/d) (2)*	*OPEC on world (percentage) (2:1)*
1970	45.7	23.4	51.2
1971	48.1	25.3	52.6
1972	50.8	27.1	53.3
1973	55.4	31.0	56.0
1974	55.7	30.7	55.1
1975	52.7	27.2	51.6
1976	57.4	30.7	53.5
1977	59.8	31.3	52.3
1978	60.2	29.8	49.5
1979	62.7	30.9	49.3
1980	59.7	26.9	45.1
1981	56.1	22.6	40.3
1982	53.2	19.0	35.7
1983	53.0	17.0	32.1
1984	54.3	16.3	30.0
1985	53.6	15.4	28.7
1986	54.8	18.3	33.4
1987	56.1	17.6	31.4
1988	58.0	19.7	34.0
1989	59.6	22.0	36.9
1990	60.4	23.3	38.6
1991	59.9	23.4	39.1
1992	60.0	24.4	40.7
1993	59.7	24.7	41.4

Source: *Oil & Gas Journal, OPEC Annual Statistics Bulletin*

The industrial countries have complex political systems, based on personal freedom and democracy; their economies are dominated by the market. The State has the power to cajole or coerce people and companies towards common goals, which have to be chosen and confirmed through a process as tentative as the market itself. The effect of official policies is ultimately defined by the reactions of private and corporate actors, who act to suit their own convenience. In these countries, therefore, economic and social events result from a myriad of individual choices by private people and companies, rather than the decisions of any one operator.

The economic and social diversification of these countries strongly influences their oil industries. At the time of the oil crisis, in many countries a State-owned

Figure 5.2 OPEC – Crude oil prices from 1970 to 1993 ($/barrel)

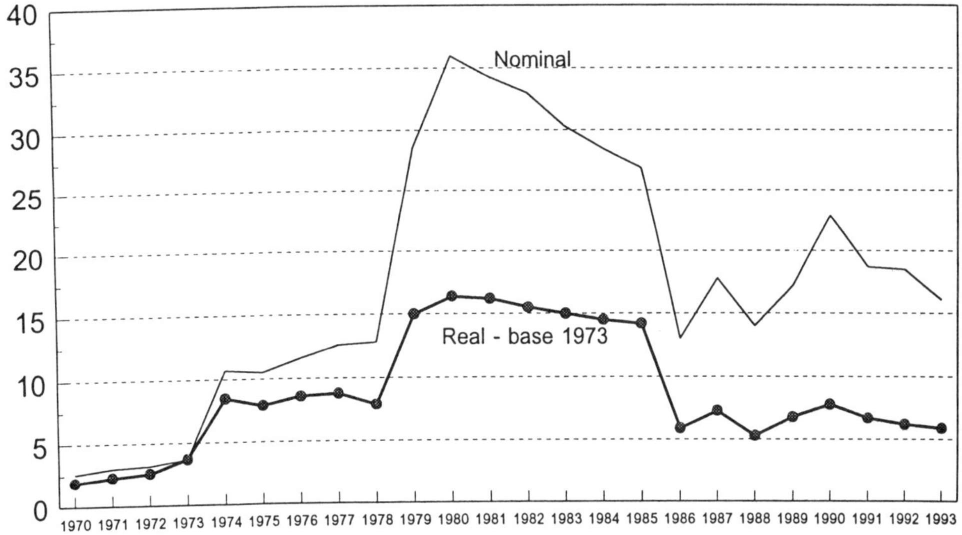

Figure 5.3 Crude oil production from 1970 to 1993

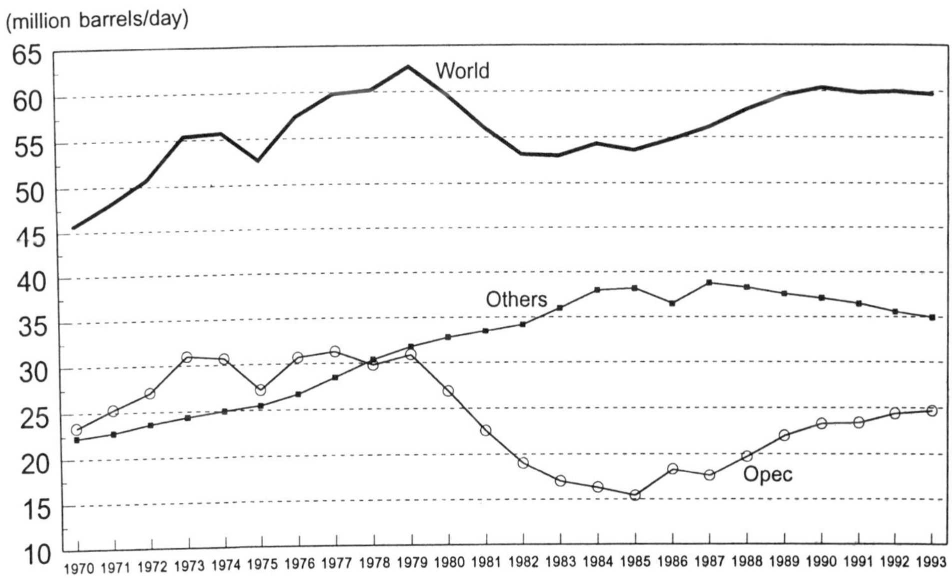

Figure 5.4 OPEC – Oil revenues and imports from 1970 to 1993

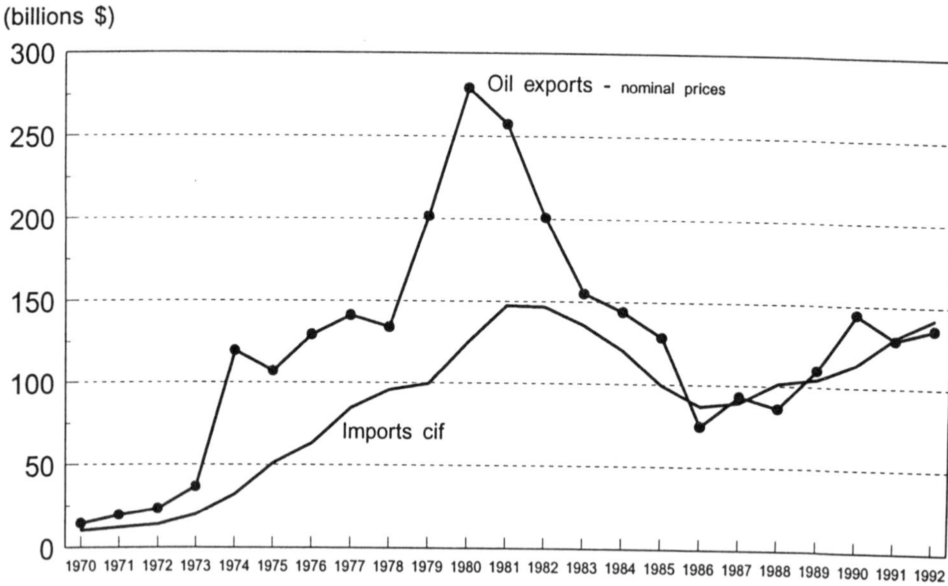

or a private oil company acted as the "national oil company" by institution or tradition. European oil companies, such as BP, Total, ENI, Elf-Erap or Veba, were – and some still are – totally or partially owned by the State and even private companies (for example, Fina) were to some extent considered as representatives – or even agents – of the "national interest" of their country. The same can be said of the American oil companies, whose interest abroad was normally identified with that of their country, and of the Japanese ones, which, although private, were – and to some extent still are – strongly co-ordinated by their government. All these companies had an important, but not decisive, say in oil matters: it was uncertain whether the government would follow or even seek their advice.

In the USA, about half the internal oil production was controlled by many relatively small "independent" companies, which exerted a stronger influence than the multinationals on the political machine. Domestic production of energy had been regulated since the Depression and the New Deal; imports of foreign oil had been regulated and restricted since the late 1950s. The oil industry was run as a tight, government-controlled oligopoly, strong enough to resist the onslaught of both the radical muckrakers and the multinationals; as a result, the oil price was higher than in Europe or in Japan.[2]

[2] For a good description of the US system before the shocks see Andrew Shonfield, *Modern Capitalism – The Changing Balance of Public and Private Power*, Oxford University Press, 1964, and particularly Appendix V – The Control of the United States Oil Market, p. 439.

The industrial world had emerged from the ruins of the Second World War (or from post-war conversion) thanks mainly to the development of the oil-based industries, which had become the leaders of the new economy. In the long post-war boom the oil industry and those directly or indirectly related to it (motor vehicles and steel, petrochemicals and plastics, etc.) had enjoyed diminishing production costs, due to large-scale production, technological progress and stable or falling oil prices. They were able to feed the market with new or improved products at declining prices. In the resulting expansion, both investment and final demand increased rapidly.

In the early 1970s, this impetus slackened. The world economy was becoming rigid and prone to overheating, and the rate of technical progress was slowing down. Millions of workers had shifted from Southern to Northern Europe and the latter was beset by overcrowding, overloaded social services and inflation. People were growing less willing to pay for the disruption caused by further development: they seemed more interested in stability and a better "quality of life" than in development, and were inclined to accept the prophecy of doom of the Club of Rome[3] that demand for labour, raw materials, oil and capital was inexorably outstripping supply. In short, the early 1970s were the tail-end of the longest period of economic growth in the developed world's history.

Small wonder that when the oil shock finally came, many people saw it as the spontaneous self-adjustment of an overheated, resource-short economy.

The tail-end of the boom had also created a rift between the consuming countries. Europe and Japan were growing faster and proving more resilient to cyclical fluctuations than the USA, who seemed weak in the basic industries and prone to "overheating", and attributed Japan's and Europe's advantage to their lower oil price.

This trans-Atlantic disparity in economic buoyancy had been one of the causes of the monetary crisis of 1971, which had freed the dollar from its 25-year-old fixed link with gold and, as a side-effect, had convinced the oil producers that an increase in prices was the only way to protect their income.

The first reactions to the jump in the price of oil were quite uncertain. The new prices seemed too high to last: many thought they would soon be reduced either by the market or by political action. While waiting for the miracle, the oil users found that they could not immediately reduce their demand and had to adjust somehow: those who sold goods or services increased their prices in order to keep buying the same amount of energy, thereby fuelling inflation, while household consumers tried to allow for higher energy cost by saving less or by reducing other expenditures, thereby depressing overall demand.

---

3 The Club of Rome, a group of authoritative experts and politicians, commissioned US universities to study depletion of natural resources worldwide. The results were very "alarmist" indeed. See Meadows, *The Limits to Growth*, a paperback by Universe Books, New York, 1972.

Unsold cars piled up in the industry's yards, with drastic effects on the suppliers, especially steel.

The increase in energy costs hit especially the energy-intensive industries. For example, the cost structure of petrochemicals was changed overnight by the sudden increase of its main raw material (virgin naphtha, a product of crude oil refining). The industry, whose costs used to be dominated by capital charges, became fully dependent on variable (i.e. raw material) costs.

Accelerated inflation and a dropping demand went together with a strong currency drain, which hit the balance of payments, especially of Europe and Japan. In this situation the rigidity of real wages left governments no choice but to slam on the brakes, using the old instruments of monetary control.

The economy stopped in its tracks and a fully-fledged recession ensued. Both demand and the utilisation of industrial capacity flopped for the first time since the early 1950s. The recession that followed, by far the worst since the 1930s, had two main components: a general decline of demand and a serious industrial crisis.

The oil-based industries, once the leaders in the world, burdened with high costs and over-capacity, cancelled their investments and began to "rationalise" operations, i.e. to shut down capacity and shed workers. The ones that had strong market power, like the automobile industry, passed on their increased costs, soon to find that higher prices further reduced demand. The new emerging leaders, the "electronic-related industries and services" were still too small to fill the gap.

All over the industrialised world, energy users tried to reduce their consumption and/or to shift to cheaper energy sources. This could be done only by investment: in a new car, in roof and window insulation, in a change of burners or boilers, in different machinery or in a brand-new productive or logistic system. However, investments take time to be decided and implemented, especially in a situation of high uncertainty.

Meanwhile, governments were trying, amid a lot of dissension, to come to grips with the energy situation proper.

The political system proved to be no faster than the market. The politicians and the bureaucrats who form the government in Western countries shared a basic culture, Keynesianism, which taught to smooth out cyclical fluctuations and to control unemployment by increasing global demand, using instruments which influence the few macro-variables that control the rate of growth, but do not tackle structural problems. Moreover, even that effectiveness had been reduced: the countries had become too rich, their economies too complex, their prices and wages too rigid for them to work properly.

Political dissension about the role of the State in the economy did not help either. Should the State step in and lead the way with inducements, subsidies, or straight orders? This question touched upon the basic balance of Western society and could not be answered on the grounds of expediency.

The European and Japanese governments controlled the bulk of electricity and coal, and owned or influenced some large oil companies; taxation there represented a hefty portion of the consumer price of gasoline and of other energy sources.

So circumstances obliged them not only to set the prices of some of the most important energy sources but also to take some longer-term measures, like diversifying the country's energy supply, or increasing domestic energy sources through nuclear energy or more exotic alternatives. They also tried to influence the behaviour of consumers through legislation, for example by limiting the hours at which they could heat their homes or drive their motorcars, or through incentives or persuasion.

The USA had options not available to the other industrial countries, thanks to their high domestic output of hydrocarbons and coal and their considerable political and military strength. These two elements made the Americans confident in their ability to "solve" the energy problem, which, however, became a hot political issue, to be treated with wide-ranging domestic and international measures, devised according to the political beliefs of the White House occupant. But political action requires consensus, and this did not come easily.

From the very start two basic schools of thought developed. The first school believed that the OPEC had really played the Last Trumpet, heralding the end of the Western economy and society. The West should change in a hurry, with the governments leading the change to reduce the hardship of transition. Although the ideas on where the transition should lead were not very clear and rather diverse, this school unconditionally supported the development of "soft", renewable sources of energy and preached conservation as a mixture of moral duty and clever engineering.

The second school wanted "business as usual". The Western economy was powerful enough to absorb the oil shock without having to change its basic structure. By inflation, energy conservation, or any other means the market would adapt to high oil prices. Governments might very well finance consumers who wanted to erect a solar jig on their roofs, but should stay well away from more serious matters, like the price and the supply of crude oil.

In the USA and in Europe these two groups were present, and quite vocal, in every party and social group, and that made the discussion confusing and sometimes misleading.

In these conditions, governments found it very hard to give clear, consistent rulings. Very often the decisions were not implemented before being swept away by opposite ones. In other cases, measures were not ready to be implemented until the market for which they had been devised was no longer there, and their effects were therefore different from, and sometimes the opposite of, what had been expected. In the USA the emphasis was first put on self-sufficiency, but soon shifted to the political and military defence of the countries supplying oil to the

American market. Huge investment programmes were launched with Federal support to devise cheap substitutes for oil, and then scrapped to reduce Federal expenditure and to encourage private investment in exploration and production of oil and gas.

This effect was visible in foreign policy, too: the USA strongly advocated the creation of the International Energy Agency as a tool for confrontation towards OPEC; while consolidating their relations with OPEC's largest producer, Saudi Arabia.

However, these difficulties did not prevent the consuming countries from putting into effect measures intended to reinforce market reactions, including fiscal measures, incentives, direct investment in energy projects, additional expenditure on research and development, etc. Instead of lowering the petrol tax to reduce transport costs, the consumer countries increased it almost to a man, claiming that it was necessary to stimulate conservation. Actually, the decision was in almost all cases motivated by the needs of an ever-larger State budget. "Energy Plans" became very fashionable on both sides of the Atlantic: their aim was to reduce both energy demand and the share supplied by oil, through conservation, development of nuclear power and renewable sources, and a greater role for coal and natural gas.

Of course, not all these measures were effective, or produced the desired results. Many projects remained a dead letter: some cost too much; some still lacked a proven technology; some were made too risky by the high increase in the prices of investment goods; others, like nuclear energy, were not accepted by the public; others again, like coal, had been neglected for too long. On the whole, however, the new approach worked, mainly through conservation.[4]

The whole thing worked better and with greater ease in the USA than in Europe or Japan. Abroad, the US oil companies were hit by nationalisation; but they were able to maintain a strong presence in Saudi Arabia, the oil exporter with the largest potential, which was becoming dependent on US political and military protection. Internally, as oil had been dearer in the USA than elsewhere, US industries were less affected by the oil shock than were their competitors, who had lost their advantage on energy costs. Moreover, the consumption of energy had been so wasteful that even drastic cuts were possible. Last but not least, the increase in crude oil prices more than doubled the value of an important domestic resource. Part of it was mopped up by Federal taxation, but a lot remained in the coffers of US "independents" and multinationals; in due course this created an exploration boom, which in turn attracted foreign capital. Energy-saving invest-

---

[4] See M. Colitti and P. Baronti, "Energy policies of industrialized countries", reprinted in *Energy* Vol. 6, pp. 233–262, 1981. This paper, presented at the First Arab Energy Conference in Abu Dhabi in March 1979, made an exhaustive review of all these measures. It ended by warning all concerned that the measures were not a charade, and that their effect on oil demand would be felt before long.

ments become profitable for small households as well as big companies, and even the sacred cow of US civilisation, the large, wasteful car, came under fire. Slowly, energy demand began to decline, just in time to be sent tumbling by the second oil shock.

The oil shocks and the measures taken by governments and consumers alike gave fresh impulse to a structural change that had been creeping during the boom years: the reduction of the economy's energy intensity is one of its main effects. The whole manufacturing, factory-based economy began to give way to a service-based, white-collared, information-processing economy, which uses less of both labour and energy.

The new leaders of the modern economy, the "electronic-related industries and services", offer a more efficient way of producing and consuming practically everything. They need little labour and low investment, and therefore have a small favourable effect on the demand for capital goods, and an adverse one on both labour and energy, which began to be replaced by new equipment, tighter control of operations, and changes in product and process design. Such far-reaching changes in the structure of the economy would have occurred in any case; however, the oil shock accelerated them, and made them happen against a background of stagnation and unemployment.

The oil-exporting countries

Immediately after the first oil shock of 1974, the producing–exporting countries looked like a looking-glass image of the consumers. Recession and pessimism prevailed in one area, development and optimism in the other. The West and Japan were going through a severe recession and a scale-down of their oil-based industry, while the exporters' oil earnings had increased by a factor of 3, and they were trying to spend a large part of it in internal development.

"Fewer and fewer people tearing up and down the 'autostrade' in beautiful Ferraris and a better life for our peasants", an Iranian Minister told the author immediately after the first oil shock. "Why should an Iranian peasant work twelve hours a day and still be very poor, while a German or an Italian works six hours and lives in splendour?"

For whatever it was worth, the outburst expressed the basic expectations of the ruling elites of many OPEC countries. Massive wealth would be transferred from rich to poor countries: the West, and especially Europe, would decline to the advantage of the oil exporters. Some people talked of an "Arab Renaissance"; others, more soberly, of an international division of resources in favour of countries which had been poor and colonised, and were now becoming rich and powerful.

The feeling had a justifiable origin in what the exporting countries had achieved *vis-à-vis* what they had perceived as their rivals, the multinational oil companies. For years the best universities in America and Europe had focussed on large enterprises[5], and had built these companies up as the main element of the international economy and the new great world power. Now the rich Goliath had been slain by a poor David: countries which had been colonies of, or at least subservient to, the rich ones had won back not only a large part of their resources, but also the power to fix prices and production levels.

That the image of the multinationals had been previously blown up out of all proportion by sycophants or simply by fashionable talk did not matter. The exporters' victory was there for all to see: the fact that it had been obtained, at least in the case of Saudi Arabia and Iran, without antagonising the USA or even the multinationals, was simply another proof of the winner's ability.

The perceived victory created perhaps too much enthusiasm among the peoples and politicians of the Middle East and Latin America. It gave rise to a picture which was too simple, and would eventually mislead many people into thinking that OPEC success would not produce a strong reaction.

This reaction started first in the world's financial system. The enormously increased dollar earnings of the exporters could not be spent immediately: their "absorption capacity" was so inadequate that a lot of resources would simply stagnate in an enormous "dollar overhang", further depressing the world economy and playing havoc with already disorderly financial markets. The oil exporters were well aware that their inability to use such large amounts of dollars meant that part of it would be recycled back, i.e. find some placement either in the USA or in the Eurodollar market. Part of the money was kept in nearly liquid form; a policy often criticised, but vindicated years later, when the countries which had done so could contain the effect of the fall of oil revenues by recalling liquid resources from abroad. Other countries chose to buy shares in famous companies and high-technology enterprises.

Meanwhile, the world's financial and banking system had to find acceptable placements for an unprecedented liquidity, while trying to keep up the level of global demand. The developed countries were in a deep recession, while the developing ones could be counted upon to turn any loan into additional demand. Further threats to the West's dominance of world affairs could be prevented by propping up countries that would otherwise collapse under the combined weight of their internal difficulties and of their oil bill. A large part of the recycled dollars ended up in loans to poor countries struggling to stay alive in the crisis. Some of

---

[5] see, for example:
- Raymond Vernon, *Storm over the Multinationals: The Real Issue*, Jan 1977, Harvard University Press
- Raymond Vernon, *Big Business and the State: Changing Relations in Western Europe*, Jan 1974, Harvard University Press
- Raymond Vernon, *Technology Factors in International Trade*, Sept 1970, National Bureau of Economic Research, Inc.

them were insolvent from the very start; some did not even try to invest their money, but simply used it to keep a level of internal consumption – and foreign placements – well beyond their means.

Very soon the game became too big and too risky, and recycling problems were superseded by questions about the repayment of these enormous debts and the resulting instability of the international financial system.

By this time the exporters had realised that they could not avoid sharing out their new wealth with the rich areas of the world. Although this gave them a good amount of power in the financial establishments, it was obviously making them more, not less, dependent on the West.

These disappointments did not cancel out the positive results of the oil revolution. The exporters were really running the upstream phase of the oil industry in their own interests, turning to their lasting advantage the biggest peaceful redistribution of resources in history. A lot of the new riches was being either transferred to their citizens as free social services and sheer hand-outs, or invested in their countries in social infrastructure and industrial plants. Many countries thought to add to the oil industry a new modern industrial sector which would, in ten to twenty years, generate enough income to offset the decline of oil revenues due to reserves depletion.

Anyone who visited some of the producing countries from the early 1970s onwards must have seen the results of these policies. Whole new cities have sprung up and a new middle class has been created, with modern tastes and inclinations. The change was particularly strong in the Arab countries, which not only "transplanted" whole industries from the West, but also imported a large mass of consumer goods. All these embodied an unmistakable Western model of life, which had little to do with the traditional Arab mores, the very ones that some of these countries openly declared their intention of defending.

Moreover, the oil exporters were forced to realise that their new oil wealth did not automatically translate into world-wide political power, or a bigger say in their respective parts of the world. Given the power, the Arab oil exporters would doubtless have established a completely new political and military scenario in their confrontation with Israel, and the Mexicans would have dealt in a completely new way with their big neighbour. But this kind of power was not to be had, and the oil exporters remained unable to solve their own vital political questions. Actually, the very area of the Middle East was soon shaken by war, as the antagonism between Iraq and Iran exploded in a war in the very heart of the oil region, deeply dividing the exporters' camp.

Iraq's war with Iran created a new and very serious security problem for the whole Gulf; the antagonism among the OPEC members ran much deeper than disagreements about oil policies could ever have done. There was very little that diplomacy could do to maintain the consensus-building process inside OPEC.

Not surprisingly, the great enthusiasm that had greeted the first oil shock did not last long. The fall in crude oil demand following the price increases of 1979–80 surprised the exporters as much as the price increase in 1974 had surprised the importers, and the former's reaction was neither quicker nor wiser. The doubling of the price in 1979 and the further 26% increase in 1980, coming on top of the first shock, did have an important effect on demand. The wisest OPEC advisers might have suspected this from the very start, but they did not seriously influence the price–quantity policies of their governments, the relationship between people in power and experts not being visibly better in the oil-exporting countries than in the oil-consuming ones.

The exporters' immediate reaction to the second oil shock was to say that prices had been pushed up not by OPEC's decision, but by panic buying by consumers alarmed by the cut in Iranian production following the revolution. Some restocking of petroleum products had in fact occurred before prices flared up; whether the multinationals anticipated an increase in prices is not clear. However, accusing the International Energy Agency of intentionally misleading them did not get the exporters off the hook: if the multinationals had indeed engineered the increase, that meant that OPEC had no control on the market; if it had resulted from spontaneous market reaction to supply–demand imbalance, then why did the fall in demand not reduce prices?

Eventually, the OPEC countries took the new prices as settled, and sat down to enjoy the multiplied dollar flow into their Treasuries; if demand continued to fall, they reasoned, they could always scale down production. If the worst came to the worst, they could always live with less, trim down some of their investments, and save oil for future generations.

So, the new prices were not only maintained, but made sacrosanct, thus creating a near-impossible problem of face for any OPEC member wanting to go back to more reasonable medium-term levels. The disagreement of Saudi Arabia could not force the price back: its decision to increase production and to keep a lower price simply created a glut, piling up stock which proved difficult to absorb. It did not, however, fail to create resentment among OPEC exporters, a persistent cause of mistrust between the largest producer and the other members.

An unexpected victim of the second oil shock was OPEC's attempt at defining a long-term price policy. A formula had been worked out by a special Long-Term Committee, chaired by the Saudi Oil Minister, Zaki Yamani, and called the "Taif Formula" after the Saudi summer capital city. Basically, it proposed the indexation of crude oil prices to the economic growth of the consumers, world inflation and the value of the dollar. The consumers might have accepted it, at least tacitly, starting with pre-1979 prices, but even the exporters realised that they would not agree to a sum total of long-term planned increases plus sudden jumps.

The OPEC countries' response to the fall in demand that followed the price increases of 1979–80 shows a disturbing symmetry with the consumers' response in 1974 and 1979. A consensus to do something new did not exist, and was too difficult to create; in such tense times even objective analysis became politically suspect. Like the consumers in 1974, when France did not join in the creation of the International Energy Agency, the producers failed to agree on a common stand, and a two-tier price system came into being which did not cancel the increases but reduced them for some privileged buyers. Apart from enabling some oil companies to make some money, it simply made the situation more confusing.

The upshot was that both exporters and importers accepted the fact that the quantity of oil produced by OPEC and consumed by Europe, Japan and the USA would be reduced by high prices. The consumers adjusted their economies through slower growth, accelerated structural change, and an increasing diversification of energy and oil sources; the exporters accepted a reduction of output and the future prospect of lower oil income and slower economic development. They may have hoped to supplement their resources with financial income from foreign investment and, in the longer term, with the new industries. Whatever the reason, the oil prices were left to exercise all their effects on demand.

The oil companies

Political action, and especially international politics, was not seen as the answer to the oil shocks. The answer was left to the market, where the oil companies operated. These had been directly hit by the crisis, and lost not only a good part of their equity oil but also the power to fix prices.

They reacted as they could, within the range of options they had. They bought less from the oil exporters because there was in fact precious little to buy, and that little had a very high price. As we have seen in the third chapter, the share of primary sales – that is of the oil bought from the oil exporters – in their total supply went down from 42.5% in 1980 to a minimum of 26.4% in 1986; while the share of their own production went up from 57.5% in 1980 to 67.2% in 1983 and 73.6% in 1986. The companies also tried to increase their production in non-OPEC areas as much as they could. Non-OPEC crude production increased by 1.1 mb/d in the two years 1974 and 1975, and 3 mb/d in the 1976 and 1977: in 1975, the year in which OPEC production was cut by 3.5 mb/d, the non-OPEC crude increased by 0.5 mb/d.

Among the companies hit by nationalisation, one, Gulf Oil, actually entered into a crisis from which it never recovered, and led to its disapperance, an extraordinary – albeit isolated – "death of a multinational". The large oil companies regrouped, and sat tight. They were short of crude, and so was everyone else, and oil demand was also going down.

However, they were facing a double crisis, a supply crisis upstream and an even

Figure 5.5 World and Middle East – Original proven oil reserves by year of discovery from 1961 to 1992

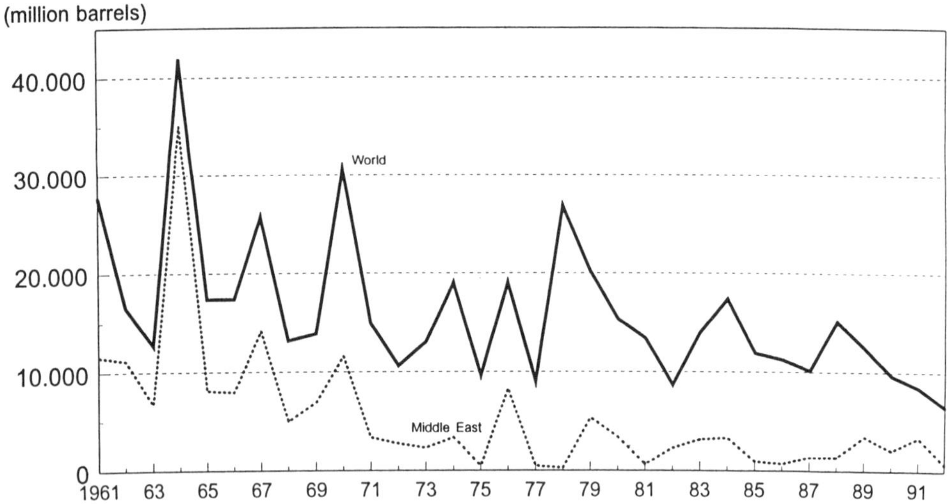

deeper one in refining and distribution. The decline of demand had all of a sudden made refining capacity excessive, and projects just completed found themselves with no market. Governments did what they could to stop the increase of the industrial prices – as opposed to the tax-paid ones – of oil products in order to control inflation, and downstream losses got very large. Some of the more tightly controlled markets, e.g.the Italian market, were abandoned by some of the multinationals (in the Italian case by BP and Shell).

In reality, the oil companies could do very little. Irrespective of what might happen to demand, they could not (and did not) react against OPEC prices; nor could they sell below cost. They had no choice but to pass on the increased prices to consumers. They used whatever freedom they had to look for cheaper oil, and to streamline the downstream system as much as they could.

Their most important decision was in fact to boost exploration, a choice which was perfectly justified by the jumps in crude prices. And they succeeded in finding new oil. Figure 5.5 shows the oil reserves discovered each year from 1961 to 1992 in the world and in the Middle East. There is a quite visible increase in non-Middle East discoveries after 1974 and, even more, after 1979.

Companies do, however, engage in politics, and the oil multinationals do so more than anybody else. If a market reaction was not on the cards, a strategic one might very well have been pursued. The multinationals were the only ones who knew the new oil lords, trained as the latter were in company-sponsored universities and colleges. They had discussed oil matters with them since the early

1950s, and understood their positions and motivations.

In fact, during the first oil shock the whole world was a-buzz with rumours about how the oil companies, prompted by the American government, had taken a "reasonable" negotiating stance *vis-à-vis* the demands of Iran in the last phase of negotiations leading up to 1974. Some of their representatives even went as far as privately saying that they had actually expected a price increase, but though they could contain to, say, $7.5 per barrel, the price at which the Middle Eastern oil could be sold in the USA without damaging the internal producers.

There is no way to know the truth of such matters. If the multinationals had an advantage in negotiating with the producers, they did not make use of it. In practice, the road not taken by exporters and consumers, and the flexible market-led nature of the oil companies, had a cumulative overall effect: to let prices exert to the full their adverse effects on the demand for oil.

An assessment

The first question that these events bring to mind is why the consumers did not react politically. They did not fight the exporters, or anyone they might have singled out as responsible for the sudden crisis. They remained apparently cool and collected during the worst period; they painfully adjusted, and neither threatened the exporters nor really tried to argue their case with them. This might be read as a favourable sign, meaning that countries no longer impose their will on others by negotiation or force; or, more simply, as lack of confidence in the ability of international institutions to solve conflicts. Extraordinarily little discussion took place between consumers and exporters and, when it did, most of the talk normally came from OPEC and OAPEC representatives, who explained their position with a rather effective mix of political arguments (basically anti-colonialism) and economic ones (mainly taken from the Club of Rome type of analysis) and concluded by saying that OPEC was doing no more than the market would have done anyway.

A big change in the world economy, involving among other things a redistribution of resources which was planetary in scale, took place without anyone feeling the need for negotiations or even discussions on a scale matching the importance of the problem. Institutions, it is true, were used: conferences were held and "new economic orders" were called for, but little of this looked like serious diplomacy. The meetings that took place were not really designed as occasions for negotiations and remained basically propaganda exercises.

Nor did the consumers approach the problem in a systematic way. The very governments that had meticulously planned relatively minor features of their countries' economies during the boom did not really try to steer the energy-induced change. The energy plans of the countries that prepared them were normally a collection of statistics on energy demand and supply, some forecasts, high or low, according to the point in time and the inclination of the compiler, and some programmes for alternative sources. Little or nothing was said about the adjustment process that was changing the economy, and even that was normally only in aggregate, macroeconomic terms.

The major effort on the consumers' side, the creation of the International Energy Agency, was at best partial. The absence of France did not really impair its operations, but although the Agency worked seriously on the possible remedies to another enforced scarcity, it could never really function as a centre to co-ordinate the transition of Western economy to a dear-energy situation. The high quality and strong commitment of its managers and personnel could not remedy this weakness, which has been embodied in the Agency's constitution and in the difficulty to reach a consensus.

At this point we have to ask ourselves: could consumers and exporters have pursued a different strategy? Was there any possibility that negotiations might have succeeded in smoothing out the problems and avoiding the sequence of shock and countershock? Speculative questions are difficult to answer. An agreement might have found a basis in the common interest of the two main actors to maintain the momentum of economic development. A good basis is, of course, not enough for a political process to start and to be seen as advantageous by two sides confronting each other. The reason why it did not happen is perhaps to be found more in both parties' general political prejudices and fears than in any aspect of the oil business. The exporters wanted all they could lay their hands on, actually feeling virtuous in their post-colonial resentment; and the consumers deeply distrusted them.

CHAPTER 6

THE OBJECTIVES AND INSTRUMENTS OF THE EXPORTERS

Management by volume, or the short-term approach

The revolution of the early 1970s made the oil-exporting countries organised in OPEC responsible for the management of world oil supply. Although their control did not extend to world production as a whole, in the early 1970s it covered a large portion of the oil produced in the world, and an even higher one of the oil internationally traded, that is, excluding from production the amounts produced and consumed in, say, the USA or the USSR. The exporters saw themselves, and the consumers saw OPEC, as the decision-maker on oil supply and prices.

What were the objectives of the new lords of the most important industry of the world?

Looking back to these years, a first tentative answer would be that such objectives, as long as they were visible, were related more to the appropriation of the industry, by necessity limited to the upstream sector, than to the running of the industry. This is equivalent to saying that they were of a short-term nature.

The exporters wanted first of all to increase their oil revenues, which had grown by $5.6 billion per year, rising from $14.5 billion in 1970 to $36.9 billion in 1973, and were supposed by almost universal consensus to be poised for a limitless increase.

This consensus was based on the generally accepted theory that oil demand is inelastic to price, because oil is a subsidiary good used in combination with others, like gasoline in a car; because oil – having no competitor – is "inevitable" to both industry and the household. Coal, supposedly the only serious competitor, had much higher production and utilization costs; all its return on the market could do was, therefore, to produce a rent in favour of oil.

The increase in exporters' revenue was to be assured by appropriating the oil rent which was beforehand – so it was maintained – divided between consumers and oil companies.

Together with that we could detect a political design, an attempt by a number of "undeveloped countries", as they were then called, to change the political balance in their favour.

The Venezuelan and Arab initiatives for the creation of OPEC and the Arab and Iranian strategists of the first shock clearly had that in mind, as was clearly

M. Colitti and C. Simeoni, Perspectives of Oil and Gas: The Road to Interdependence, 91–102.

shown by the partially failed embargo against the "allies" of Irael against Egypt. The oil exporters' basic instrument, the very foundation of their policy, was the nationalisation of the oil reserves.

Without nationalisation, and the consequent change in structure of the oil business, crude prices could easily have gone up, at least to equal the level at which crude was sold in the USA.; but very probably the exporters thought that the oil companies would not – or could not – keep a very high crude price for any length of time, being too committed to the West's economy to risk being seen as enemies. In any case, they would have kept a large part of such increase themselves. The exporters believed that there was an "oil rent", and felt entitled to it.

The nationalisation changed the structure of the industry, which became even more concentrated than before. Its previous oligopolistic character, due to its concentration in few very large integrated oil companies, was reinforced by the organisation of the exporters in a common body, with the intention and the means to become the ultimate controller of supply. The new institutional structure – which had a good basis in the geographical concentration of oil reserves – had one fundamentally new feature. The oil exporters could not follow the multinationals down the integration route: they could gain control only of what was in their territory. Refineries and markets remained well outside the exporters' reach, and the industry acquired the "looped" character which it has not yet lost. Gone also were the interlocking concessions, substituted by a link among the exporters, their membership of OPEC. It is difficult to resist the temptation to anticipate that the new link would eventually prove to be weaker than the previous one.

The nationalisation of oil reserves made it possible for the exporters to tailor the level of production and of supply to the objectives they had on the price of crude. Crude production was reduced both from the newly nationalised fields and by reducing the lifting rights of the oil companies when the latter still operated.

In this way the exporters could fix the price by a decision which every exporting country took in autonomy, but which was discussed and approved in the OPEC meetings, the venues where decisions on prices and quantities were publicly taken.

The management of crude volumes needed to make crude prices "stick" had important effects on the structure of the oil market. Supply contracts between the exporters and their lifters were revised downwards for quantity and upwards for price so repeatedly that they lost their significance in favour of the spot market, on which the exporters offered, at very high prices, marginal quantities of their crude. This was in fact the instrument and the result of the short-term approach: from 1979 onwards it enormously increased its share of oil transactions. This was a development with a vengeance, as the spot market will, of course, accept any crude offered for sale, to become in time a good channel for new non-OPEC sources of supply, which deprived OPEC of its ability to fix the overall supply ceiling.

In these years, when crude was scarce, the spot prices were systematically higher than the contract ones. They measured in fact the value of marginal quantities which the exporters did not want to sell through contracts with the oil companies, and sold instead on a different market and at a higher price: and their quotations were used to put upward pressure on the official selling price.

These complex policies eliminated for exporters (or so they thought) any need for communication with the consumers of their product. Apart from commercial communication, the two sides had little to say to each other, and spoke to justify their activities to public opinion rather than to try to understand each other.

The effects of such a strategy were very important, all the more so because some of them were quite unexpected. The producing countries succeeded in their main objective, and had an astonishing increase in their oil income. However, a number of relevant negative effects were also produced.

On the supply side, their policies changed the composition of supply, that is, which oil fields were produced first, and which were the marginal ones, or, to use words of that time, who was the "swing producer".

On the demand side, it basically reduced it. We will now discuss these points, starting with a critique of the argument about the appropriation of the oil rent. The reader will need to wait for the final chapter of this part to read a corresponding critique of the exporters' basic idea on demand elasticity.

The oil rent

The price–quantity policy of the exporters was intended to secure for them, as they themselves said, the whole of the "oil rent". This is a key element which has to be looked at in some detail.

There are at least two kinds of rent, the so called "consumer's rent", and the "mining rent". The first of these the consumer enjoys when he pays for a certain good or service less than he would have been willing to pay; i.e. when the price the consumer pays is lower than the "value" he attributes to the good or service he needs. To support this somewhat hazy concept with some historical evidence, one might think that in the years immediately prior to the first oil shock the consumers had perhaps been prepared to pay up to the $7–8 per barrel paid in the USA rather than the $3–4 per barrel paid in Europe for the Middle East crude. After the 1974 increase, which went well beyond that, oil demand stagnated for only one year – in which the market "digested" the shock and the embargo – and then started growing again, although at a slower rate. This may be taken as an indication that the price resulting from the first oil shock was not really preposterous, but simply cancelled out the consumer's rent, and what was difficult to accept was really the manner of the increase – the shock and the embargo.

A second kind of rent, which we can call "the mining rent" comes from the difference between the cost of the cheaper source of some commodity and its price, which will be set by the market on the cost of the marginal – and costlier – supply. If the amount of cheap oil which can be produced cannot satisfy total demand, then other, costlier oil has to be produced and put on the market at a price which must cover its cost, which is the marginal cost of supply. This will give to the cheaper producer a rent equal to the difference between his production costs and the market price. If the underground resources are the property of the State, it will appropriate this rent through taxation and/or State-owned companies.

A third concept, the "energy rent" ensues from cost differentials among different energy sources. The numerical example given in Table 6.1 – in which coal represents all non-oil energy sources and competes with fuel oil – should shed some light on this concept.

Part A of Table 6.1 includes the production cost – including some "normal" rate of profit – of different qualities of oil and coal imported into Europe and/or produced in Europe, European coal being dearer. These estimates are not far from reality, and the argument would remain valid even if we took these figures simply as numerical examples.

Part B calculates the rents created by different prices of oil. In the first column of B_1 the price of oil is equal to the cost of the dearest source of oil: i.e. \$10/b, the cost of oil C of Part A. Oil A and oil B appropriate all the rent coming from the cost differentials among the various oil sources; coal does not appear on the market, because the cost of production of its cheaper quality is higher than the dearer oil.

The cheapest oil, A, has a big rent; oil B has a much smaller one, while oil C, although on the market, has no rent: it is simply sold at cost to satisfy the last portion of demand. As the level of rent per barrel also measures the ability of a source to compete with the others, at this price oil A is in a very good competitive position, and its production may increase to satisfy rising demand.

If the oil price goes above the cost of marginal oil supply, it will first readmit coal on the market, and then give some rent to the cheaper variety of it.

In the second column of B_1 and B_2 oil is priced at \$12/b, and coal D enters the market, but with no rent, while all types of oil have one, although diminishing with the increase of cost of production. The competitive position of the best oil worsens, because even the costlier oil has some rent and, therefore, an incentive to production.

However, the price of oil can be increased beyond the marginal cost of coal.

In the last columns of B_1 oil is priced at \$30/b, which becomes \$27 per barrel of fuel oil. This price is higher than the marginal cost of both oil (\$10/b) and coal

Table 6.1
COST OF PRODUCTION OF OIL AND COAL, AND RENTS ACCRUING TO BOTH AT VARIOUS PRICES OF CRUDE OIL AND FUEL OIL
(dollars per barrel)

A: Cost of production

Oil		*Coal*	
Oil A	3	Coal D	12
Oil B	7	Coal E	18
Oil C	10	Coal F	21

B: Rents accruing to oil and coal
B_1: *The "oil rent": cheap oil versus dear oil*

Crude oil prices: *Rents*	*10*	*12*	*15*	*18*	*30*
Oil A	7	9	12	15	27
Oil B	3	5	8	11	23
Oil C	No rent	2	5	8	20

B_2: *The "energy rent" fuel oil versus coal*

Fuel Oil Prices: *Rent:*	*10*	*12*	*15*	*16*	*27*
Coal D	–	No rent	3	4	15
Coal E	–	–	–	–	9
Coal F	–	–	–	–	6

– Not on the market

Notes:

Costs include a profit margin.

Cost of coal. Type D: South African at $57/t; Type E: Australian–Colombian at $84–88/t; Type F: USA at $105.5/t.

From dollars per ton of coal to $/b the price is divided by 0.69. The result is increased 30% for additional use and discharge costs.

Price of fuel oil equals 80% of the crude price plus $2-3 for transport cost.

($21 per barrel of oil equivalent).

As a consequence, all crudes and all coals enjoy a rent, even the costlier coal having $6 per barrel equivalent. This has two main effects. The cheaper oil loses its competitive advantage *vis-à-vis* the dearer oil (the latter having a rent not much lower than the cheaper one); and all oils lose their competitive advantage *vis-à-vis* coal, which can be marketed aggressively. Oil is therefore substituted quickly and widely.

This is what happened during the 1980s, when the cheapest oil (of OPEC origin) was gradually pushed out of the market by non-OPEC high-cost oil, and coal got back into the market.

A price higher than the marginal cost of supply of both energy sources will in fact make the consumer indifferent towards them; he will tend to reduce his dependency upon the source with the highest share of the market – that is, oil – to strengthen his defences against possible future increases. In this way the oil exporter's hope of appropriating the "energy rent" is disappointed.

In conclusion, the producer may very well appropriate the oil rent, with prices in line with the cost of the dearest oil required to satisfy demand. And he can appropriate the "energy rent" only if he prices his oil at the marginal cost of the best coal, and not at the "marginal cost of coal". He cannot follow the latter criteria without reducing the competitiveness of his oil as well as his share of the market.

It is not surprising, therefore, that new production capacity was created after 1974 not only in oil, but also in coal and natural gas, which found easy ways into the market and went directly to reduce oil's share.

The composition of supply

The high price policy of OPEC and its control over the level of production did give a strong incentive to continue or start production in non-OPEC areas.

High prices of crude oil encourage exploration and development, as the expenses and capital outlays are offset by expectations of high revenues for as long as the life of the fields.

However, high prices depress demand and new oil discovered and developed at a high cost may not find any market. This is a serious deterrent, especially for big projects which reach the break-even point only at high rates of production. On the contrary, low prices stimulate demand and expand the market to accommodate new oil. However, they do not give rise to profit expectations from projects of high-cost oil; and are an even more formidable deterrent.

Stuck between the devil and the deep blue sea, the oil companies preferred high prices, considering that their main income comes from crude oil production, while the downstream sector – which normally suffers most from high crude prices – is no more than a necessary evil to dispose of the crude; and that selling non-OPEC crude at near-OPEC prices would in any case result in profit. The companies did their own free riding, together with non-OPEC producers: given OPEC's high prices, elementary logic told them to find ways of substituting OPEC oil with their own.

OPEC kept prices up, and got the blame; while the oil companies refined profitably costly oil that would never have gained such a market share if it had to compete with Arabian oil. Exploration and production were energetically pursued outside OPEC, even in the USA, the area where the chance of making a big cheap find was – and still is – the lowest, but where the profit per barrel found was and is fairly high.

Non-OPEC producers were all the time able to ride the OPEC prices, and to increase their market share, while OPEC's was consistently reduced, together with the rate of depletion of its reserves. Non-OPEC producers pursued a policy of depleting their fields to the maximum economic rate (MER); but the reserves being so intensively depleted were not the cheapest ones, but rather the opposite.

The technical cost of producing oil is a complex matter. The cost of oil includes the cost of preliminary exploration, i.e. the activities prior to the first drilling, including general surveys and studies, geology, and all indirect – e.g. seismic – exploration; and the exploratory wells. All these are normally treated fully or in part in the oil companies' balance sheets as expenses and not investment, which is tantamount to saying that they are amortised in the year of completion.

When a field has been drilled, and the find is proven "commercial", the oil company must put it in condition to produce the amounts required by drilling development and production wells and installing the necessary capital equipment (gathering lines, producing platforms, etc.). How many wells have to be drilled depends on the size of the field, on the possible production per well and on how much the company wants to produce. The development of an oil field is a "bankable" enterprise, and it is often done with project financing: therefore the rate of interest, and the intricacies of the financial agreement, greatly influence the cost.

Given this complexity, a breakdown of world reserves according to the technical cost per barrel produced would be rather difficult, and require details beyond our scope. However, qualitative parameters can be used with little fear of producing a radically wrong sequence. The bigger the field and the higher the unassisted production per well, the lower the technical cost; the older the field, the lighter the financial cost; *ceteris paribus*, offshore oil costs a lot more than onshore oil.

According to these parameters, the huge onshore fields exploited since the 1940s – or earlier – in the Middle East countries rank among the cheapest in the world. These are, however, the very ones that in the ten years after the shock were exploited at very low rates, while the recent and relatively small offshore fields of, say, the North Sea, were depleted at the maximum possible rate.

In short, high-cost oil displaced low-cost oil, and the cheapest reserves were produced only after everyone else had sold his product on the market. The oil world would in any case be a far cry from the Ricardian model (developed, admittedly, for agriculture), according to which the best fields are produced first, and only an increase in demand can call into the market the less favoured fields which, when they come in, give "rents" to the owners of the best ones. Exactly the opposite was happening at that time in the oil industry.

The inversion of the priority in depletion did seriously affect the exporters: it was perhaps the main cause for the crisis and the collapse of their policy.

The priority inversion meant not only that non-OPEC reserves were produced at maximum rate, while cheaper OPEC reserves were produced at a much lower rate; it also created the same effect among OPEC producers. The OPEC producer with the largest and cheapest reserves of all, Saudi Arabia, became a sort of "producer of last instance", that is, it took over the role of maintaining other OPEC members more or less at their agreed quotas, "swinging" its production with demand; reducing it when demand decreased, and, in theory, increasing it if and when demand went up.

The latter was a purely theoretical case, and Saudi Arabia saw its output progressively decreasing until it represented a fraction of its capacity. The fall of production went step by step with the fall in oil revenues.

This uneasy "allocation system" went on, until Saudi Arabia's patience snapped and the largest oil exporters in the world changed strategy.

Was there a way to avoid this development, or was it an inevitable consequence of OPEC policies?

When OPEC oil started losing ground *vis-à-vis* non-OPEC oil, some exporters concluded that a variant of the "management by volume" strategy might successfully apply if an agreement could be reached among all the exporters, OPEC's members or not, on the level of supply. It is clearly a kind of "total monopoly" dream, given the difficulty of accommodating in a single framework the interests of many different producers, some of whom enjoyed for many years a profitable free ride on the others. However, apart from the formidable political obstacles such a project would create, the strength of competition from other energy sources would ensure that in the medium-term the conclusion would be the same. Oil demand would decline even if a high price policy were to be followed by all the exporters without exception; what is more a "total monopoly" of supply would reinforce the tendency to fix high prices. The attempt to obtain all the energy rent for the oil producer would in fact make oil even more vulnerable, in the present, and in the future, to competition from other sources of energy.

The energy market and the consumer countries' economies

On the demand side, OPEC prices prompted a strong reaction by the consumers, which brought to life an element up to then dormant: the energy market of the consuming countries became very competitive. Fuel oil, coal, natural gas and some nuclear electricity competed for thermal bulk uses in industry and in the utilities; gas, diesel oil, LPG, nuclear electricity and some coal competed for household consumption. Even automotive fuels, long believed to be without substitutes, gradually came to be menaced by competition from octane boosters and oxygenated products. Oil was losing its share of the market to its competitors across the board.

Moreover, oil also had to compete with capital. With high oil prices, investments to reduce industrial and household energy use often had a pay-back period of mere months, and were put into effect despite the high rate of interest. To quote but one example: virgin naphtha-based steam-crackers were designed and built in the early 1970s with an overall energy consumption of 8,000–9,000 kcal per kilogram of ethylene produced; after 1979, engineering firms were offering new plants which consumed as little as 5,000 or even 4,500 kcal per kilogram of ethylene produced. The much-reduced fuel consumption of the car engine is another example, often quoted because of its dramatic improvement in efficiency.

Legal and institutional restrictions on energy consumption were introduced to limit the freedom of consumers, especially in the home. The production mix in industrial countries changed rapidly. Services and electronics, which consume little energy, became the leading forces in the modern economy, occupying the place once held by the oil-related industries.

Some of these trends were lagged responses to the price shocks; some were not. All together, they built in the energy market mechanism for curtailing demand, especially for oil.

All this was achieved by running down the downstream industry, and by investing heavily in new equipment. Refineries and oil-related logistic systems were shut down: the depressive effect of overcapacity on product prices was so strong that the refiners eventually forgot their habit of keeping plants in operation even if they cover only the marginal cost. Big investments went to nuclear energy and other sources, as well as to their logistics, which promised cheaper energy or, perhaps, simply a supply not dependent on OPEC.

The final consequence of this policy was an important change in the outlook of the oil producer.

The time-horizon of the producers

The main oil exporters, i.e. the OPEC members with the lowest costs, the swing producers, did seriously reduce the rate of depletion of their reserves, thereby lengthening their time-horizon. Some of their oil will be produced later than originally expected. Is this in the producer's interest? The answer is not easy, because economic theory becomes quite hazy whenever time is considered, and also because the issue involves non-economic preferences. Is it better to use the available resources now or to leave a large part of them to future generations? The mood of the 1970s suggested an answer that seemed based on a kind of disdain of present plenty rather than on economic reasoning.

Strict economic methodology does not give exhaustive replies on this point. It usually discusses the value of a dollar now and in the future, and does answer that a dollar now is worth more, not only for its immediate psychological value but also because a dollar will produce interest all the time from now to the point we are considering in the future. The discount rate and the interest rate play an important role in these matters. So does the inflation rate, since the real rate of interest becomes zero or negative in periods of high inflation, thereby increasing the preference for the present *vis-à-vis* the future. However, the value of a dollar cannot simulate the decision that an oil producer has to take. A famous article written in the 1930s[1] by a welfare economist, followed, but not greatly improved, by a rather rich literature, implies that, at least in the case of monopoly, the producer from non-renewable reserves can renounce his preference to producing now rather than in the future if his selling price increases over a period with whatever indicator of future value is taken – in the original article the rate of interest: in other words the producer can reach the point of indifference between the present and the future only if the steady expected increases in prices promise to "keep the value" of its sales.

The economist's tool box offers another instrument, the marginal utility of money. For an exporter with an oil income bigger than he can dispose of, additional quantities of this revenue have a very low value; he may postpone them, in the hope that they will be more valuable in the future. However, many OPEC exporters were not in such a position, as they were hit by a serious reduction in revenues already in the early 1980s.

---

[1] "The economic of exhaustible resources" by Harold Hotelling in the *Journal of Political Economy*, April 1931. See also "Hotelling economics of exhaustible resources, fifty years later", by S. Devarajan and A.C.Fisher in the *Journal of Economic Literature*, March 1981. The case discussed by Hotelling was based on metals, which are not, strictly speaking, similar to oil. Metal objects can be melted and the metal in them reused. This makes metal a sort of "permanent stock" which can only be increased. Instead, oil is non-renewable; once burnt, it cannot be recuperated.

Finally, the risk factor has to be considered. The more you postpone a certain event, the more risk you take that it may not take place, or that it may happen in a way very different from the one you hoped and planned for. In the case of oil, risk is related not only to market competition but, above all, to demand. Technology is the dominant factor, which in a decade can change the outlook for any product, even one as important and with as many uses as oil. New and improved technologies for the production and consumption of energy and new competitors may very well convert the greatest riches into well-nigh unusable products. This has happened to coal and could still happen to oil.

Summing up

From 1974 onwards, the value of petroleum exports of OPEC countries went up by $159.5 billion until 1980, the year of the maximum revenue ($279.3 billion); that is, it increased $ 22.8 billion per year, the highest jump ($82.9 billion) taking place in 1974.

Since this peak it went rapidly down and reached a minimum of $75.0 billion in 1986, losing to the tune of $34.0 billion per year, the maximum loss ($56.3 billion) in 1982.

The oil revenue went up again in 1987, and 1988, to overtake the $100 billion mark again in 1989. The 1974–1980 increase in oil revenues was due almost totally to the prices, given that production was actually lower in 1980 than in 1973. The decrease was due to both factors: production in 1986 was 8.6 million b/d inferior to that of 1980, and prices were $11.6 per barrel in 1976 against $36.0 in 1980.

In fact, the short-term management of OPEC ended up by failing on both counts; that is, on both volumes and prices.

To conclude: the OPEC strategy to manage the oil reserves can reasonably be called "short-term management" because it was concerned above all with immediate gains; and it resulted in stronger competitors – the non-OPEC exporters – who increased their production steadily slowing down only after 1997; in a reduced demand for oil due to the competition of other energy sources, mainly gas and coal; and in a decrease of the use of energy per dollar of GNP in the developed countries.

Eventually, it met with the final obstacle, the refusal of the "swing" producer to keep swinging, as this policy was clearly leading nowhere, and the larger OPEC exporter found itself producing less than 2 million barrels per day, a third of what it was known to be able to produce. Saudi Arabia had been critical in words and in deeds of the policy pursued after 1980, and had tried to adopt a different policy as far as prices and volumes were concerned. However, the two-tier prices could not work properly, and had to be abandoned.

So, the largest exporter in OPEC decided to "keep its market open" in future years, and to do so completely reversed its previous policy.

What followed was the countershock of 1986 and the following years.

CHAPTER 7

THE COUNTERSHOCK

The reasons for the countershock should be clear from what we have said up to now, so we may concentrate on the effects of the decision, which was basically Saudi Arabia's, to align their crude to the "market", and to sell at net-back prices instead of at official selling prices (OSP).

The "net-back prices" are related to the final value of the crude, that is, the price of the oil products on the consumer countries' markets.

Suppose a barrel of crude oil is refined in the following proportions dictated by the market: gasolines (20%), gas oils (40%), and the rest heavy fuel oil (a poor yield, but not very far from what was the average hydroskimming European refinery). The barrel's value will be a weighted average – 20% by the gasoline price; 40% by the gas oil price, and 40% by the price of heavy residual.

In this way a price strictly related to the market was reintroduced, with the only problem of a relatively limited transparency.

When the Saudis said "market" in 1976 they meant something different from what the exporting countries had meant in 1980, when "market" had been defined by the production and marketing policy of the OPEC members, which had kept crude oil in short supply. Six years later, "market" meant the price resulting from a myriad of transactions, almost all of them speculative, in which brokers, banks, and oil companies entered to obtain or to deliver certain amounts of crude at a certain date for a certain price. A great number of these transactions, and especially those of the companies, were "defensive" in scope: they were intended to dampen or cancel the risk of a bigger – or smaller, as the case may be – price compared with the one contracted for. Two crudes, with the less visible addition of Dubai's, were involved in these transactions: the Brent, a part of the North Sea production, and the West Texas Intermediate (WTI), a Texas crude with a purely local market.

These two crudes represented about 1% of total world production, but nevertheless became "representative" of the market because they were produced by oil companies which depleted their fields on purely technical criteria, that is, at or near MER. No production or marketing "policies" influenced the supply of these crudes, produced in countries free from the obligations resulting from OPEC membership.[1] In fact, whenever someone hints that one or more of the exporting

[1] See "Who fixes the price?" by F.C. Cirio, in *AGIP Review*, no. 26, January–March 1993, a short but effective article amid a very large literature.

M. Colitti and C. Simeoni, Perspectives of Oil and Gas: The Road to Interdependence, 103–110.

countries also create a futures market on their own crude, the answer of the market operator is always negative: supply would not be free from strategic consideration; the exporter could not be fully trusted to deliver the crude at conditions which did not suit him.

Both these crudes had a "natural" market, the United States, where the gasoline-rich Brent expresses all its value, and WTI has a large number of long time-takers.

When Saudi Arabia's Oil Minister, Zaki Yamani, said that the country would sell at net-backs, and not at OSP, he really meant that it would follow Brent quotations. In due time net-backs were abandoned, and the quotations of these two crudes became the price effectively ruling the crude oil market.

The basic intention of the Saudi Minister was to stop the loss of market for Saudi oil. OPEC oil did not represent more that 28.7% of total world production in 1985, and Saudi Arabia, the "swinging producer" was in some months not producing more than 2 million barrels a day, getting an oil revenue which was too low for its investments and social income transfers. The Saudi leadership felt that the Kingdom was heading towards a very dangerous crisis, and that anything should be done to avoid it.

For a long time OPEC had been divided into two groups, a system which had its roots not only in political divisions, but also in the difference between the volume of crude oil reserves of the various members.

The high prices of the short-term OPEC policy represented perfectly the logical strategy of the members, for instance Algeria, with limited hydrocarbon reserves.

At the other extreme Saudi Arabia, the desert Kingdom, has enormous reserves, and a small population. Little surprise that its "natural strategy" should be exactly the opposite of the Algerian. Keep the market open, defend your market share, try to create the conditions for an orderly and continuous depletion of your reserves, until well beyond the century.

This difference between "the haves and the have nots" has been with OPEC since its inception: but Saudi Arabia, politically isolated, was for a long time content with following a middle route until the OPEC short-term policy could finally be seen as directly damaging its national interest. In the mid-1980s, Saudi Arabia was fighting for its survival, or so its leadership thought, and it had to force the issue in one way or another.

The Saudi decision caused an uproar, which however did not last very long. OPEC production started picking up as early as 1986, a year in which it gained 2.9 mb/d, and, apart from a 0.7 mb/d reduction in 1987, it retained the increase: from the minimum of 1985 (15.4 mb/d), OPEC oil gained slightly more than 1 mb per year, reaching 24.7 mb/d in 1993. Its share of world production grew from 28.7% in 1985 to 41.4% in 1993. Moreover, prices did not stay as low as they had fallen in 1986 ($13.0/b): they gained $4.8/b the next year, went to a peak of $22.9/b in 1990, while oscillating between a bottom range of $14/b and an average of about $17/b, to reach $16.1/b in 1993.

However, the new course taken by the exporters was also due to a deep change in the implicit – that is, not much publicised – strategy of many of the main exporters. They had been redefining their crude oil marketing policy, and had wisely decided to integrate downstream, that is, to acquire independent outlets, refineries and distribution networks.

Going downstream the oil exporter becomes a fully integrated operator. Of course, finding space in the oil products' market is not easy: the oil companies are well entrenched and downstream integration may be costly for the newcomer, in terms of both investment and initial market losses.

The downstream has very rarely been profitable *per se*. It has always been an area in which a part of the upstream profits was sacrificed in order to sell the crude. The great illusion of the 1960s, when the oversupply of cheap crude fostered many independent companies which processed the marginal barrels of the majors and sold the products, was dispelled in the early 1970s. Only the integrated survive. This makes life hard for the majors themselves, which have to buy at least 40% of their crude from producing countries, and impossible to non-integrated refiners and sellers who, by their own very existence, create the product glut which kills them.

The idea of going back to the integrated industry is not new: it was perhaps born on the same day on which the producing countries obtained control of the oil reserves but not of the oil market. These countries themselves are now moving in the right direction, led perhaps not so much by the abstract logic of the argument, but rather by their inability to resist the competition from non-OPEC oil. Having the strongest interest in high prices, OPEC producers are unable to resist competitors who accept a lower price: in order to stop losing market share they have no other way than to integrate downstream, in order to protect at least part of their oil from the crude-to-crude competition.

Up to now the moves have been neither dramatic nor negligible. Among OPEC members, Abu Dhabi, Iran, Kuwait, Libya, Saudi Arabia and Venezuela have built and acquired interests in refineries inside and outside their countries, with a throughput of 7.8 mb/d (see Table 7.1).

As these countries in 1994 produced 18.8 mb/d, their level of integration is about 42%.

If we consider the total capacity of the plants in which OPEC countries invested outside their national boundaries (see Table 7.2) we see a total of 4.9 mb/d. Their share in them, strictly speaking, would be of 2.8 mb/d, i.e. 57.2% of total capacity; in 1994 these countries supplied these refineries with 2.6 mb/d.

Outside OPEC (see Tables 7.3 and 7.4) some important producers like China, Mexico and Norway have reached an even higher level of integration at home and abroad.

Let us remind ourselves that the companies themselves are integrated by 60–

Table 7.1
INTEGRATION OF SOME OPEC COUNTRIES IN THE YEAR 1994
(thousand barrels per day)

Producing country	*Crude oil production (1)*	*Refinery throughputs*			*Percentage of Integration level of OPEC countries (3) = (2):(1)*
		Domestic	*Abroad*	*Total (2)*	
Abu Dhabi	1,800	190	170	360	20
Iran	3,671	1,170	20	1,190	32
Kuwait	1,690	750	135	885	52
Libya	1,361	340	180	520	38
Saudi Arabia	7,907	1,650	910	2,560	32
Venezuela	2,335	1,160	1,135	2,295	98
Total	18,764	5,260	2,550	7,810	42

Sources: ENI DataBank "Cycloil"; Petrostrategies; PIW; annual reports of oil companies.

70%. Clearly, producers who have gone so far downstream must look at the crude oil in a completely different way: what they increasingly sell is not crude but oil products, and their knowledge of the real oil market, the one on which consumers operate, becomes much more sophisticated.

The main political change of the "countershock" has been the new position of Saudi Arabia. With a political decision which was by no means obvious, and that in fact, broke a habit which had been consolidating since the 1970s, the Kingdom has taken the role of supply leader. The Kingdom assumed responsibility for security and continuity of supply, and let anyone know that investments were not a problem: if and when the market required new production capacity, the Kingdom would muster the capital and the technical expertise to increase it, and to maintain the overall supply for as long as it was needed at the level the market required; and it was well understood that, although the Kingdom did not take any commitment on prices, this very decision on supply would guarantee a more or less steady price, subject, of course, to the consequences of unforeseeable world crises.

Inside OPEC some reactions were very harsh. The countries with a smaller reserve base, which have no interest in a long-term policy, predictably opened the war of words by repeating that volumes would not increase, and oil revenues would stagnate or decrease. In fact, as we have seen, oil revenues and OPEC's share of world oil production were on the increase very soon after 1986: and non-OPEC production slowed down its steady increase which had been going on

Table 7.2
OPEC PRODUCING COUNTRIES: REFINING CAPACITY ABROAD AND OIL SUPPLIED
(in thousand barrels/day)

Producing Country	*Refinery location*		*Total installed capacity*	*Owned quota (%)*	*Owned capacity*	*Oil supply*
Abu Dhabi	Austria: Schwechat		210	19.6	41	40
	Germany: Burghausen		72	19.6	14	40
	Spain :	Algeciras	170	10.0	17	20
		Tenerife	100	10.0	10	60
		Huelva	90	10.0	9	10
	Total		*642*	*14.2*	*91*	*170*
Iran	India: Madras		131	24.5	32	20
Kuwait	Denmark: Stignaes		60	100.0	60	60
	Italy: Naples*		96	100.0	96	–
	Netherlands: Rotterdam		75	100.0	75	75
	Total		*231*	*100.0*	*231*	*135*
Libya	Germany: Harbourg		78	66	51	20
	Greece		90	14	13	10
	Italy: Cremona		95	100	95	100
	Switzerland: Collombey		72	100	72	70
	Total		*335*	*69.0*	*231*	*180*
Saudi Arabia	Korea: Onsan		146	35.0	51	145
	Philippines: Lymay		147	40.0	59	145
	United States:	Convent	220	50.0	110	
		Delaware	150	50.0	75	
		Port Arthur	255	50.0	128	
	Total		*918*	*46.1*	*423*	*910*
Venezuela	Bahamas: Freeport		500	100.0	500	–
	Belgium: Antwerp		15	50.0	8	10
	Curacao†		320	100.0	320	200
	Germany:	Gelsenkirchen	227	50.0	114	75
		Karlsruhe	152	16.5	25	30
		Neustadt	144	12.5	18	20
		Schwedt	230	18.5	43	45
	Sweden:	Goteborg	12	50.0	6	6
		Ninasham	28	50.0	14	24
	UK:	Dundee	10	50.0	5	10
		Eastham	22	25.0	6	22
	USA:	Corpus Christi	165	100.0	165	160
		Houston	265	30.0	80	135
		Lake Charles	320	100.0	320	190
		Lemont	150	50.0	75	140
		Savannah	30	100.0	30	28
		Seaview	85	100.0	85	40
	Total		*2,675*	*67.8*	*1,814*	*1,135*
Grand Total			*4,932*	*57.2*	*2,822*	*2,550*

Source: ENI DataBank "Cycloil"; Petrostrategies; PIW; annual reports of oil companies; Petroleum Economics Ltd.
*Refinery closed in September 1993; it is used as a product storage facility.
†Leased

Table 7.3
OTHER PRODUCING COUNTRIES: REFINING CAPACITY ABROAD AND OIL SUPPLIED
(in thousand barrels/day)

Producing Country	*Refinery location*	*Total installed capacity*	*Owned quota (%)*	*Owned capacity*	*Oil supply*
Cina	US: Pacific	55	50.0	28	20
Mexico	Spain: Sommorostro	240	34.5	83	200
	USA:	216	50.0	108	110
	Cuba: Cienfuegos	140	*	*	65
	Total	*596*	*41.9†*	*191†*	*375*
Norway	Denmark	71	100.0	71	70
	Grand Total	*722*	*49.8†*	*290†*	*465*

* Not published
† Excluding Cienfuegos

Table 7.4
INTEGRATION OF OTHER PRODUCING COUNTRIES IN THE YEAR 1994
(in thousand barrels/day)

Producing country	*Crude oil production (1)*	*Refinery throughputs*			*Percentage of integration* level *(3) = (2):(1)*
		Domestic	*Abroad*	*Total (2)*	
Cina	2,950	2,570	20	2,590	88
Mexico	2,673	1,440	375	1,815	68
Norway	2,499	274	70	344	14
Total	*8,122*	*4,284*	*465*	*4,749*	*58*

Sources: ENI Data Bank "Cycloil"; Petrostrategies; PIW; annual reports of oil companies; *Oil & Gas Journal.*

unimpaired since 1970, had the first check in 1986 and, after a little rebound in 1987, went on a decline again. The demand elasticities and the competitive advantages were again working in favour of OPEC. The point was that the Kingdom of Saudi Arabia, and not OPEC, was driving the dance, and even the drama of the Gulf War, and the short-lasted $40 per barrel, did not ruin the new set-up.

The consumers, for their part, felt that the whole problem of oil supply was now in more credible hands, and that the long-term approach chosen by Saudi Arabia would actually work. After 1980, the peak year for the exporters, when OPEC's volumes, prices and incomes started to decline, the oil importers had actually lost interest in the oil scene. The oil policy of OPEC was no more a real threat and oil had gradually gone off the front pages. Europe was going through a great diversification move, the shift to natural gas, preferred for its cleaner quality, and also propelled by technological innovations that made it very competitive for the production of electricity.

The countershock of 1986 was felt by the West as an inevitable conclusion of a painful experience that everyone wanted to forget as soon as possible. Many problems were still open, but the feeling was not of problem solving: the West would rather be happy than worried.

The Saudi decision finally gave the West the idea that a cycle had been closed, and that the "market" had taken over the oil industry again. The market had in fact taken over, but if speculation was more acceptable than before, it was still risky; and many, including some of the older and most respected industrial and banking companies in the West, took just too much of it and fell victim of unpredictable development or of foolhardy behaviour.

The countershock was only the conclusion of a period in which the oil exporters saw their position steadily worsening. However, it did affect non-OPEC producers, too, and especially the larger of them, the USSR, which had pinned its hopes of a steady development of its economy, and of the welfare of its citizens, on an increase in the price of oil.

Exports of oil from the USSR to the West had started in earnest in 1960, and had developed through long-time contacts with large buyers, ENI–AGIP first of all. In the 1970s, the exports of oil had been supplemented by the exports of gas, and by the early 1970s the USSR was experiencing a very comfortable hydrocarbon income, which was multiplied by the oil shocks. For the first time the economic management of the USSR was seeing a flow of dollars entering the country which could reduce, and perhaps solve, the long-standing problem of the balance of payments with the West.

The oil shock was confirming to the economic and political authorities of the USSR that the recipe for economic development was in reality quite simple.

The large investments in the production of oil and gas, which had already started, were promising a strong increase in energy exports and in dollar revenues. The bet seemed to bear little risk, and the forecasts of future crude prices which the USSR economists and engineers took from their Western counterparts were extremely buoyant: \$40/b, perhaps \$50/b in 1990, and more. In this way the higher levels of consumption and of imports from the West which had started in the late 1960s and early 1970s could be maintained without any effort and, above all, without an overhaul of the obsolete planning mechanism and production technologies.

Perhaps a few figures will be sufficient to show both the reasons for such wild hopes, and the size of the disappointment. The USSR in 1992 exported 1.8 mb/d; at the price of \$40–50/b that would have produced a total of \$26–32 billion per year; however, at the price the USSR actually received (about \$18/b) the revenue was less than expected by \$14–20 billions per year.

The investment had already been done, however, and the rest of the USSR economy – the military section apart – had been badly starved of capital until it could not function properly.

The economic, and later political, collapse of the USSR is partly due, I believe, to this miscalculation. Its leaders thought it could free-ride on OPEC prices, while at the same time actively contributing to the erosion of OPEC's market share not only with increased exports but also with new, fast-developing exports of natural gas. The Soviet system was unable to withstand the consequences of the miscalculation: it was by that time extremely rigid, and had been led for so many years by people not willing to improve their structure of the economy, while playing the great world power. The abandonment of that role (the "peace dividend" as it came to be called in the West) was not enough, because by the the USSR had run out of time.

CHAPTER 8

SUMMING UP THE GREAT CHANGE

Introduction

The full circle of the oil industry from the early 1970s to the mid-1980s now requires a sort of final wrap-up. We will therefore attempt first to use the economist's tool box to interpret the changes of the physical and economic quantities, and of the elasticity of demand and its time frames. We will tackle the political angle in the final part of the chapter, with a brief historical excursus on the oil industry. A summing up will follow, which will try to open the way to the second part of the book.

An economist's interpretation

Some graphs

The years from 1970 to 1987 can be divided into five different periods: the avant-shock, from 1970 to 1973; the two shocks, from 1974 to 1979; the crisis, from 1980 to 1982; the slow pick-up, from 1983 to 1985; the countershock, 1986 and 1987; and the years up to 1994.

The main data are presented in synthetic form in Table 8.1. They are also presented in the following figures.

At the start of the first period, in 1970, oil was \$2.6/b, a price which probably gave the oil customers some rent. Up to 1973, both demand and prices increased, which can be interpreted as the consumers' rent being reduced but not cancelled by the increase in prices. From 1974 to 1979, oil prices exploded twice, in the first and in the last year, and kept increasing in between. In the whole period, prices rose from \$10.7/b to \$28.6/b; OPEC production oscillated, reaching its all-time peak in 1977, with 31.3 mb/d, and ending the period with a very slight increase, while world production kept a high rate of increase, reaching its all-time peak of 62.7 Mb/d in 1979.

It seems clear that although the 1974 increase had wiped out the consumers' rent, the price of \$10.7/b was not really "punitive" for the consumers who, in any case, could not reduce their oil demand overnight. Oil production was pulled along by the lack of alternatives, as the exporters thought it would be.

The second shock changed that: in the third period oil demand reeled under the

M. Colitti and C. Simeoni, Perspectives of Oil and Gas: The Road to Interdependence, 111–124.

Table 8.1
MAIN DATA FOR THE FIVE PERIODS FROM 1970 TO 1987

	Oil Production (mb/d)		
Periods	*OPEC*	*Total*	*Oil price ($/b)*
I: 1970–1973			
1970	23.4	45.7	2.6
1973	31.0	55.4	3.7
Differences	7.6	9.7	1.1
Average annual rate of increase	9.8%	6.6%	9.2%
II: 1974–1979			
1974	30.7	55.7	10.7
1979	30.9	62.7	28.6
Differences	0.2	7.0	25.3
Average annual rate of increase	0.1%	2.4%	21.0%
III: 1980–1982			
1980	26.9	59.7	36.0
1982	19.0	53.2	33.0
Differences	–7.9	–6.5	–3.0
Average annual rate of increase	–	–	–
IV: 1983–1985			
1983	17.0	53.0	30.3
1985	15.4	53.6	26.9
Differences	–1.6	0.6	–3.4
Average annual rate of increase	–	0.5	–
V: 1986–1987			
1986	18.3	54.8	13.0
1987	17.6	56.1	17.8
Differences	–0.7	1.3	4.8
Average annual rate of increase	–	2.3	36.9

impact, and fell by 6.5 mb/d; OPEC production fell by 7.9 mb/d. The first year of the period, 1980, was the peak year for prices, as 1979 was for demand.

In the fourth period, demand for OPEC oil kept falling slowly, while total oil demand slightly increased. The period started at a lower price level – $30.3/b – and kept falling.

The last period starts in 1986 with the "countershock": prices fell (from $26.9/b to $13.0/b) and demand for OPEC oil jumped 2.9 mb/d, the first increase since 1979.

Figure 8.1 World oil production and nominal prices from 1970 to 1993

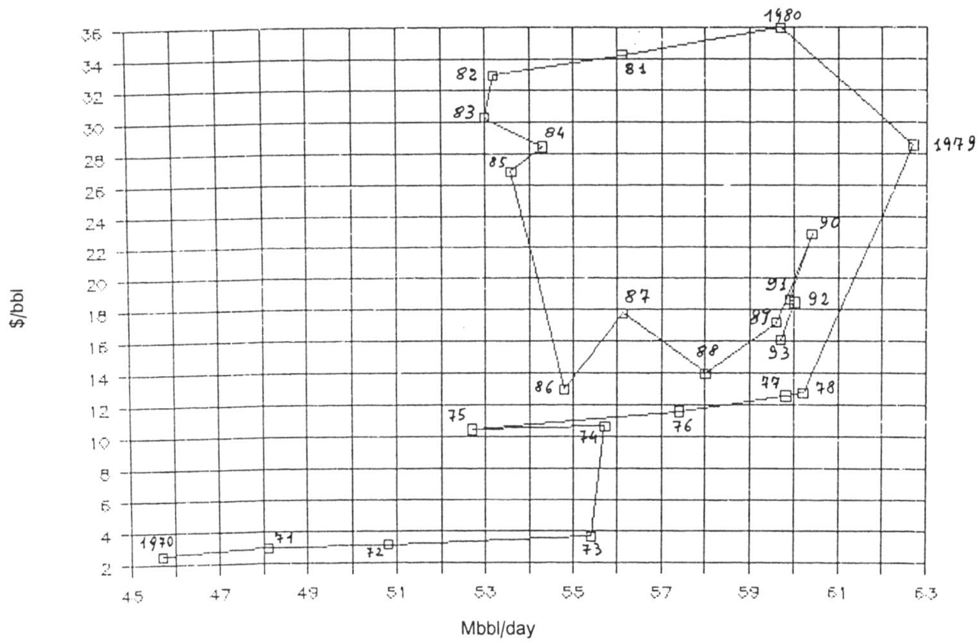

If we take the first four periods together (1970–1985), to exclude the after-shock, we may draw three curves for the three main phases: low price, high demand increase; increased price, slow demand growth; high price, decline of demand. Figures 8.1 and 8.2 present these curves, the second drawn with the very simple technique of linear regression. Can we put them together in a single curve? Figures 8.3 and 8.4 represent the application of the parabolic model, with a 95% confidence limit, to the world and to the OPEC data. The interpretation does not change. Demand and price rise together until they reach a turning point, after which demand decreases sharply with any increase in prices.

We may, however, pay more attention to the last period, considering the figures for 1986 which represent another drastic price change, this time downwards. The increase in demand is much more visible in Figure 8.5, with lines similar to those of Figure 8.2. If we try to express all the five periods with one curve we find that the parabola is no longer enough, because demand does not move along the upper arm of this curve, but starts to increase lower down, producing a sort of ripple, or curl, on top of the curve.

The statistician's panoply is so rich that we can find a model, a clothoid, which can be used to represent this phenomenon. The clothoid curve is a sort of spiral, and is also called "Cornus spiral" (Figures 8.6 and 8.7; see also Annex 3). This curve tends asymptotically to a value, which can be calculated to $23.0/b in

Figure 8.2 World oil production and nominal prices (linear OLS, annual values from 1970 to 1985)

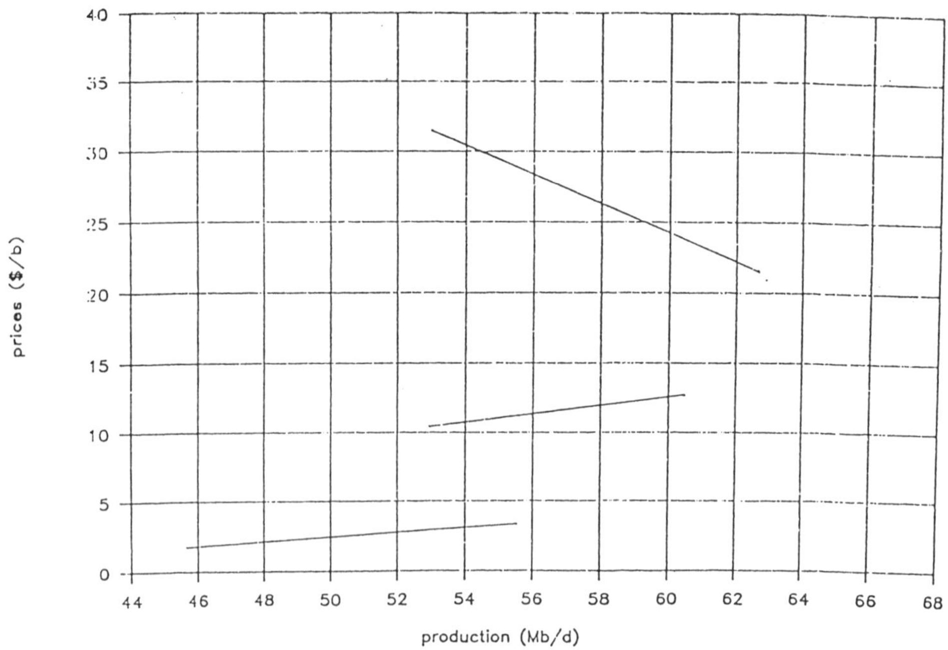

Figure 8.3 World oil production and nominal prices (parabolic model, annual values from 1970 to 1985)

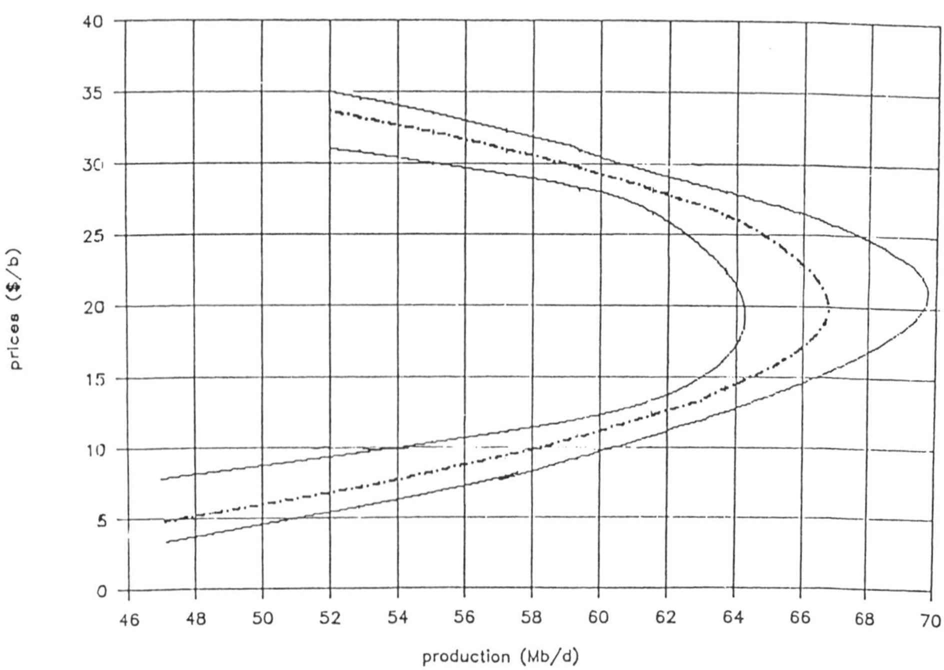

Figure 8.4 OPEC oil production and nominal prices (parabolic model, annual values from 1970 to 1985)

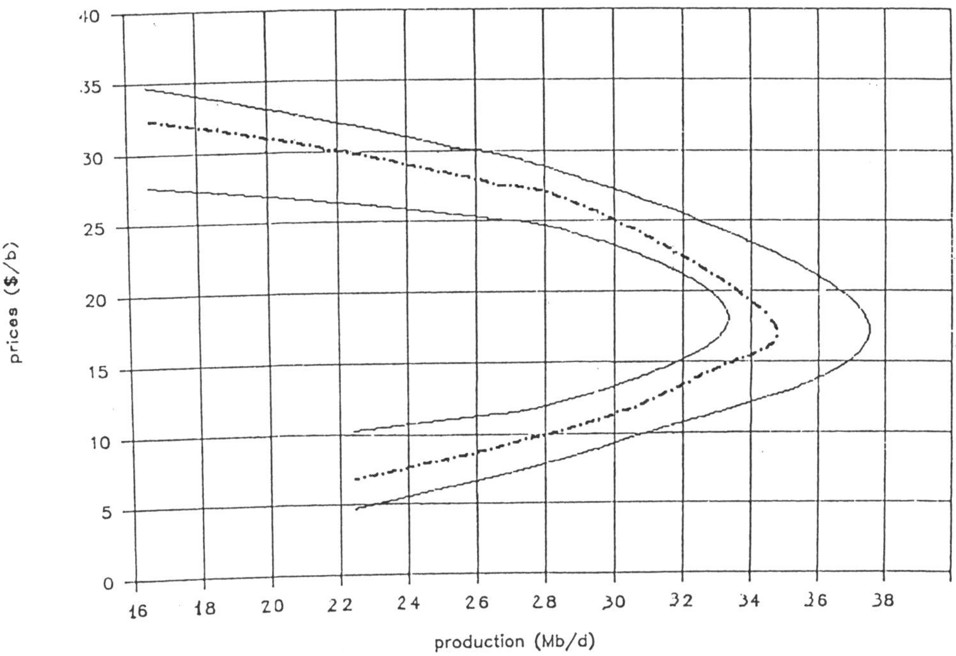

current dollars with a world production of 56 mb/d. This might be considered the point at which the market would tend in the whole period. The curve now looks like a bass clef, as those of the first batch looked like half an egg. What do these curves mean?

They represent the historical development, and their mathematical fit is good: but it is difficult to say whether they also represent the "law" of change. They certainly do not seem to be usable for forecasting prices and demand. For this, curves must be built on some continuity, while ours are exactly the opposite. They point out discontinuities, and you can never be sure that the next one will have the same shape or the same effect as the last. Neo-classical economists would consider these graphs as absurd: according to them one level of demand is compatible with more than one price; in the final graph, with three.

What clearly comes out of this analysis is the importance of time-lag in the behaviour of the oil consumer. The consumer's reaction to price changes take a long time to have any practical consequences. In order to influence his oil consumption he will have to change his car, insulate the roof of his house, change the factory's boiler or its complete energy system.

Oil is not like tea or coffee, and its demand is influenced not by changes in consumer taste, but by structural shifts: by differences in the rate of growth

Figure 8.5 World oil production and nominal prices (linear OLS, annual values from 1970 to 1987)

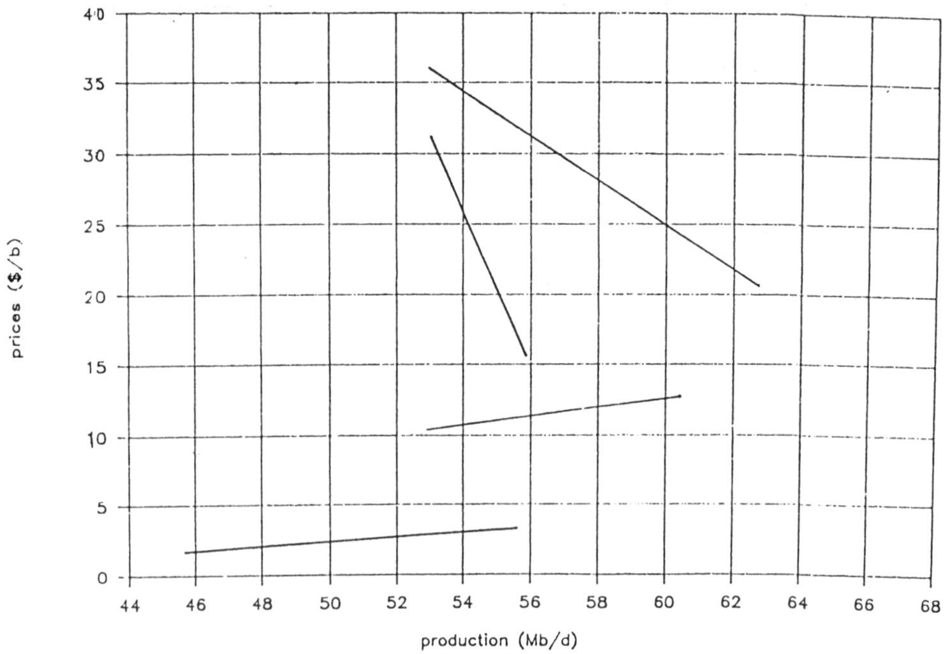

Figure 8.6 World oil production and nominal prices (clothoid model, annual values from 1970 to 1993)

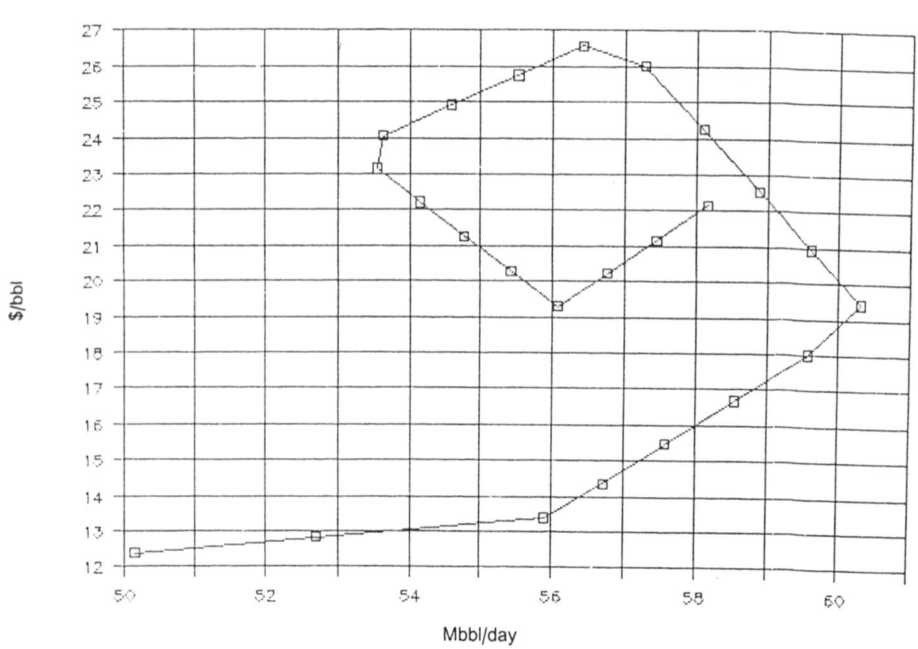

Figure 8.7 World oil production and nominal prices (clothoid model □ annual values from 1970 to 1993); historical data (+)

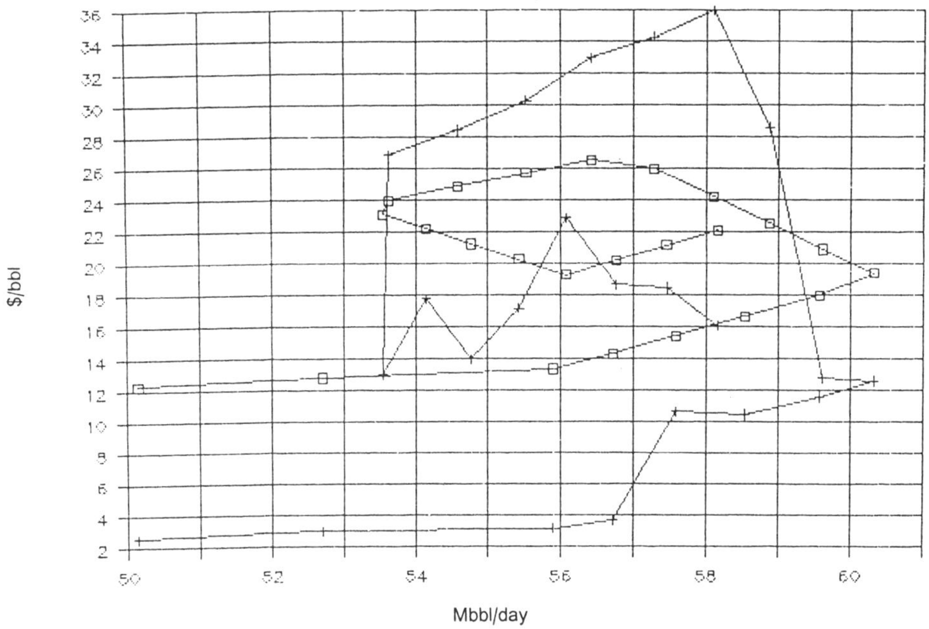

between services and industry, and between heavy and light industries; by changes in the pattern of population on the territory due to higher transport costs, etc.

Demand starts changing only a while after the price increase, but its changes are cumulative, and therefore produce an acceleration effect. At the time of the 1974 shock the oil exporters thought that demand was inelastic, and that consumers could not reduce their oil consumption. Having taken oil production into their own hands they felt they had absolute power, which would last as long as they stuck together. They were right, but only for a time. Initially, the consumers had no power over the oil market. However, the passing of time and the second shock induced them to build up a market power by direct control over their own demand and by creating *ad-hoc* institutions, which tend to remain active well after the disappearance of the original stimulus. A high level of confidence is needed to persuade consumers to abandon their defences once they have built them up.

These, in short, are the reasons why the egg-shaped curve gave way to the bass clef, and strange new instruments risk being added to the panoply of the statistician. If the curves look absurd, the reason may be that they represent an absurd reality.

A Cyclical Model

It may be argued that, like any other product, oil has a "cycle" of the interactions of demand and supply, and that the decisions of sellers and buyers work only as activators.

Suppose that a new product comes on to the market, offering better services at a lower price than products already in use. Its use spreads in new areas and by taking over existing uses from the old products which are gradually relegated to the margin of the market. When the "new" product's market share is high enough, its price will increase to reflect what we may (improperly) call its monopoly value – and perhaps its scarcity value, if supply, being managed by the producers, has not developed quickly enough. High prices will call the old products back to the market, or create some new ones, and the demand for the dominant product will stop growing. The relative market shares will become better balanced, only to become unbalanced again if and when the cycle repeats itself.

This is perhaps what could happen to oil. Coal was the only real industrial and heating fuel until hydro-electricity and oil began to weaken its position. Oil made its really big entry in the European economy in the 1940s when it was abundant and cheap, and also much more flexible and easy to use than its competitors. In the post-war boom it grew in its own uses (transport), through the creation of a new industry (petrochemicals), and by substituting coal in steam raising and heating. Its price remained stable up to the early 1960s and then fell in a period of aggressive selling by the oil companies which by the early 1970s brought oil's share to 55% of total world energy. Oil had become the most important commodity in the world.

Then the oil price soared, and its substitution became the objective of a great number of projects, from large, high-technology ones to small day-to-day technical fixes. Such changes are not instantaneous, but are deep and lasting. Modern technology changes continually, but it is cumulative, at least in a given time-span: "technical fixes" applied under pressures keep working for many years after the original stimulus. Oil-saving techniques and equipment, incorporated in the existing capital structures, will keep working for a long time, almost independent of prices.

The model must, however, take into account the existence of two sources of oil: the OPEC exporters, the price leaders; and the non-OPEC, the prices takers and free-riders. The latter sold their oil at prices in line with OPEC ones; but their crudes were favoured because they brought profit to the oil companies which produced them. Between the mid-1970s and mid-1980s the share of coal and of the other competing energy sources did not rise as expected because it was non-OPEC oil, and not them, which competed with OPEC oil.

In short, a strong increase in the market share of OPEC oil led to an increase in prices which in turn reduced demand until prices were lowered again.

It took about half a century for oil to acquire a high market share in a much-increased total consumption. But such an exalted position could not last more than ten years before the full realisation of its "monopoly value" made it vulnerable to increased efficiency, and to that part of the crude supply which escaped the control of the price leaders.

The political angle

These non-mutually exclusive attempts to explain the oil shocks and counter-shock, and to locate them correctly in the historical development of the world's economy must be completed with the political angle. Oil is a political commodity, because it is vital to the modern economy, and to anyone who wants to create one; and because it is in large portion produced in countries which for a long time had little will of their own, and whose role was at best that of a pawn in a chess board, in which the strong pieces were in the hands of powerful corporations and colonial governments.

In order to discuss the political angle we will run quickly through the history of the oil industry, which is an important part of modern history: far from being marginal and "technical", it is always near to the core of the matter. Throughout its complex and sometimes chaotic life,some durable elements stand out, stemming directly from its very birth: its strongly impressed birthmarks.

The birthmarks of the industry

The oil industry was born in a country, the USA, and in a period, the years straddling the nineteenth and the twentieth centuries, which deeply and permanently influenced its nature, leaving strong imprints on its body and soul. Some of these are political, some are operational.

First, the development of the oil industry was from the very start identified with the rise of the USA. Oil was developed there at a time when the USA was overtaking Europe. It created new opportunities (for example, as a chemical feedstock) and gave the USA an additional innovative edge over its competitor, Europe, whose economy was based on coal, and could not switch fast enough. Whatever effort the Europeans made after the first decade of the twentieth century, the bulk of the oil industry had been American from the start, and the American interest always had a lot of weight, if not the last word.

Second, the oil industry was born in the age of colonialism, when Europe and the USA created in their respective areas of influence direct or indirect colonies, and organised them into a system which seemed bound to last for ever. Whether the oil companies simply followed the colonial armies, and took advantage of the resulting favourable conditions, or actively instigated colonial expansion, is now a moot point. They were soon identified with colonial regimes and were singled out by the colonised as its mainstay. Eventually, they took all the blame, even the part which should have been allocated to others. Colonisation also meant attrition among the colonial powers, which was very visible in the oil industry. The American companies, which had found a lot of oil near home, fought hard to subtract from Europe – which had found no oil in its African dependencies – the only colonial or semicolonial area in which it had been found, the Near and Middle East. One of the instruments of the fight was the bigger share (the so called "fifty–fifty") offered by the US companies to the local elites, a seed of the system's demise. In the 1940s and 1950s, American vocal anticolonialism did not wash the colonialist dye out of the oil industry, which was the main reason for the revolution of the 1970s.

Third, the oil industry was born in a period when national states were playing an increasing economic role, being at once the sponsors of colonial ventures abroad, the purveyors of capital, and the controllers of last resort of the domestic market. This was to prove one of the most lasting characteristics of all, in tune with the general development of state intervention in the economy in the twentieth century.

Fourth, the oil industry was born in the age of industrial integration and giantism. In those years industries agglomerated, forming colossal structures which integrated all the steps from the raw material to the finished product. This was partly due to the personal quest for power of nineteenth-century entrepreneurs, and to the discovery that the big guys could influence the market, while the small ones had to follow it, taking the good with the bad. It was also due to the restrictions of international trade in raw materials, which made supply from abroad an uncertain matter, and provoked defensive reactions. In the oil industry, integration and giantism played at their maximum strength, and influenced the industry so much that you were either big and integrated, or you were not.

To sum up, its birthmarks identified the oil industry with Western, and especially American, capitalism, in its phase of state intervention and colonialism, and of large, powerful corporations. Not surprisingly, a hostile reaction was set afoot, both in the West and elsewhere, which eventually came near to blowing up the whole structure.

The reaction

The first shot of this reaction, the October Revolution in Russia, sealed off the large resources of one of the largest producing areas in the world. It also concentrated the energies of the late Russian Empire to internal development, so that the rest of the world oil market was left to Western companies. From the end of the First World War onwards, hostility began to generalize, going from Mexican nationalisation to the creation of the state-owned companies and State control of refining and marketing in Europe. During this period, however, the strength of the large integrated oil companies reached its peak, together with the concentration and integration of the whole industry. The Second World War did not shake it, but marked the end of colonialism, thus setting in motion an all-important process of change. The creation of OPEC in 1960, and the diminishing control of the large multinationals over the crude oil supply, simply prepared the sea change which came about with the nationalisations of the early 1970s and the Iranian Revolution a few years later.

Considering the structure of the industry after this long process, we may conclude that its birthmarks, while not erased, have by now been greatly reduced, and are not nearly as important as they used to be. Revolution resulted in a partial transformation; new features were added to the old ones, which did not disappear, but remained, although diminished in importance and/or changed in meaning.

Colonialism, the element which made the whole structure unacceptable, has gone for good, and the relations between Europe, the USA and the producing countries – good or bad as they may become – will not revert to a generalised colonial domination. US dominance of the industry has been reduced by the rise of the producers and the reduced aggressiveness of the US companies abroad. The integration link has been largely severed and is being reconstructed by the exporters themselves. However, the last of the birthmarks, the importance of state control, has been enhanced, as the governments of the new countries have not let their oil companies run their side of the oil industry, but decided to do it themselves, thereby largely increasing the political element in the industry.

One might sum up the whole process of reaction by saying that it put the political approach of a group of exporting countries with big oil reserves and a strong need for economic development where the business-like mentality and behaviour of the oil corporations had prevailed. Although the latter's behaviour was certainly not devoid of political elements, and the political choices of exporters are not without their business justification, we can summarise the oil revolution fairly by saying that somewhat moved the oil industry from business to politics.

However, after the countershock, the producers found themselves at a crossroads. They could take stock of the unwanted effects of their "management by volume" policy and change it, adopting some version of a long-term policy; or they could try to continue as before because of the difficulty in reaching a new consensus. In this second case, the oil market would have remained the clashing point of contradictory interests and strategies; the exporters, the consumers and the oil companies fighting each other for money and market share; the consumers defending themselves by tight control over oil demand; the oil companies taking whatever opportunities the market would offer and adapting to shrinking demand with a reduced geographical spread. We might have a succession of soft and tight markets: what would have happened to demand is anyone's guess. My guess is that it would have remained stagnant. Probably such a non-solution would have been particularly negative for the interest of the long-term, resource-rich exporters.

Instead, the exporters have brought themselves to a change of strategy and to a long-term policy. They realised that the "management by volume" policy had landed them in an impossible situation, and that a new policy is necessary, based on a more business-like attitude, i.e. on the recognition that the market has its rules, and the consumers their share of power.

A summing up

The experience up to the present time

We have experienced at least three ways of managing the oil supply.

The first was by the multinational companies, which made a relatively good job of it (certainly good for them). They integrated from top to bottom, and shared among themselves the decisions and the risk of oil producing.

This method, efficient as it may have been, grew more and more out of step with the political environment, both in the West, where new companies clamoured to enter the club, and in the world at large, where the poorer peoples wanted a share in the riches.

The shift from the first to the second method of controlling the supply has been quite shocking. We have described it at length, not only because with the flowing of time it gets more and more distant from the present, but also because it can be seen as a case study: how the world economy and politics react to a sudden and strong increase of a very important price.

The system run by the oil exporters was based on higher prices, and an OPEC volume policy consistent with them.

Why this policy was chosen is perfectly understandable. OPEC members were not only relatively new at the game, but also restricted to one side of the industry, and had very little knowledge of the other side. The priority they attached to immediate oil income was perfectly justified, at the beginning of the '70s, by the needs of their economies and their ambitious development plans. Moreover, the accepted wisdom about the rigidity of oil demand had been proved true by the small price increases of the '50s, and no previous experience existed of price jumps similar to those of 1974 and 1979. Finally, some exporters outside the Gulf often took the chance to pre-empt decisions, leaving OPEC no alternative but to follow.

Seen in retrospect, we may perhaps consider that the experience of the "management by volume" policy was unavoidable. Inevitably, the price of crude oil came to be seen as the main indicator of the exporters' political weight, and was therefore loaded with political, and even personal, elements, which made it almost untouchable. This enormously reduced the ability of OPEC and its members to correct their course after this policy began to have a negative impact on oil demand, and to stimulate competition from non-OPEC oil as well as other energy sources.

This method of managing the oil industry very soon hit the usual obstacles a near-monopoly meets in the modern economy (potential competition becoming actual, decline of demand, previously non-competitive substitutes which become so) and ended up with a strong "countershock" which had some fearful effects – some of which were quite unexpected – especially on some of the exporters themselves.

The only practical alternative, the long-term policy, was seen as a proposal to put to the customers, and was actually sketched after much study as the "Taif formula"; it was, however, swamped by the second oil shock, and could not be revived after that.

After that came the countershock, which inaugurated a system which we can call "mixed". The largest OPEC exporter has embraced a "long-term policy", that is, a policy of depletion of its own reserves which takes into account the need to keep its marked share, and to maintain the market for its oil "in being" for a long time to come; and has more or less left to a highly speculative futures market the task of fixing the price. To put it more simply, the largest exporter has decided that it is useless to try to regulate both quantities and prices, and has settled for the first – that is, for securing for itself a large share of the market at prices at which the cheaper producer gets his rent, but very little goes to the dearer one, and none to costlier energy sources.

The system has withstood strong political and economic pressures, and it seems that by incorporating features of both the previous systems it has become relatively stable.

As we have seen, oil demand is growing, although at a slower rate, the volume of dollars going into some of the producers' coffers is by no means negligible, the economy has gone back to its cyclical flow, which is a built-in feature which cannot be cancelled.

Shall we therefore stay content, and declare the present one the best of all possible worlds? Certainly not. In any case, it is necessary to understand better what we have called a "long-term strategy" of the oil producer, also in order to see whether it offers interesting opportunities on a wider basis; second, we must have a better understanding of the conditions of stability of the present system, which seems based on a very unstable (actually, on the most unstable thing of all) speculative market.

Finally, we must answer a more general question. Is the present structure of the oil industry able to offer some solutions to the political problems which still loom very large at the world's horizon, some of which are especially relevant for the oil-producing areas?

We have up to now discussed in parallel the developments of exporting and consuming countries alike, as they both depend, although with a different intensity, on the very same industry. We seem to have taken this parallel for granted, and from the historical record we may very well conclude that the consumers have a strong say in the politics and in the economies of oil, and that whoever ignores it does it at his own risk.

We might conclude that producers and consumers are interdependent, and they have to move together, to manage not only the oil industry, but also the great political problem which it affects.

Perhaps the relative calm of the present market might allow us to consider whether this interdependence has a meaning which goes beyond the simple statement that what one exports the other has to buy.

CHAPTER 9

A MODEST PROPOSAL: A POLICY OF INTERDEPENDENCE THE NEED FOR CONSUMER CONSENSUS

Introduction

The successful completion of the oil revolution did change the oil world in more than one way. The most important one was the large amount of resources accruing to the exporting countries, some of which became centres of attraction of every possible sort of merchandise, as well as financial operators with a big clout. For the first time in economic history, the European and American industry found a large outlet in areas which had never been of any consequence before. Very soon after the oil shocks, the demand of the oil countries became very important for capital and consumer goods, the demand for which was low due to the economic crisis. Part of the money the industrial countries had to spend on oil was being converted in additional demand; to the flow of resources paying for oil another flow was added, going in the opposite direction, which increased very quickly. Both flows were large, and some people came to see in that interchange a hint of a possible solution of the exporter–importer contraposition, which was then polarising international politics. Some kind of agreement was perhaps possible, due to the fact that the two sets of countries were interdependent: the exporter could not sell oil to any party different from the oil consumers, and the latter would have to continue buying from them; in addition, both could expect to gain from a stable organisation of the other flow of goods and money which went in the direction opposite to that of oil and oil money.

The industrialized countries had an obvious interest in selling back to the oil-rich countries a part of their industrial production, and the latter in buying at reasonable conditions. Interdependence became a political and economic concept to be carefully studied and evaluated.

The first Interdependence study

In Europe and in the producing countries, in the 1970s and 1980s, some people were convinced that interdependence between oil exporters and importers was too strong a reality – and an opportunity – to be ignored; and that perhaps a pump-priming action could start the process towards rational oil and development policies.

M. Colitti and C. Simeoni, Perspectives of Oil and Gas: The Road to Interdependence, 124–135.

The discussions held by such people at the Oxford Energy Policy Club, in Rome and elsewhere in Europe, and in Kuwait, at OAPEC's premises, gave rise to a large project intended to demonstrate the reality and importance of interdependence, the opportunities it offered for a rational set-up of the oil industry, and the feasibility and merits of a development policy based on it. Studies and discussions went along for years, and a mixed economists' team worked in Rome and Kuwait for more than two years on a large econometric model, the Interdependence Model. This work was presented in 1981, in Rome, to a large meeting in Palazzo Barberini between OAPEC and its member countries and some governments, oil companies and experts from Italy and other countries in Southern Europe; and in 1983, in Kuwait, where a revision and a development of the work were discussed and evaluated.

What were the main tenets of this work, and what did it demonstrate? The whole work runs to many volumes, and cannot be exhaustively presented here; but a short review might be useful, as that opportunity still exists and the advantages to be reaped are actually greater now than in the mid-1980s.

The most interesting aspect of this retrospective presentation is that the policy it envisaged has been implemented at least partially. The main trend of events after the early 1980s was in the direction indicated by the model, although it perhaps happened in a different way. The basic tenets of the work were, in brief:

1. The oil exporters have to adopt a long-term policy, in order to defend their market share against competitors who are not to be excluded from the market. This policy has a good factual basis in the abundance of oil reserves in certain countries, and the long life generally foreseen for the hydrocarbon economy. A long-term policy includes the need to recognize the fact that the oil market cannot be run without a high degree of consumer consensus.

2. Oil exporters and importers are linked by the strong two-way trade between Europe and the Middle East, made up at the start by oil on one side and industrial and agricultural products on the other, but gradually becoming more complex, as oil exporters develop their economies and offer new products. In 1983 European exports to Arab countries were three times bigger than the exports of the USA and Japan to these countries: Arab exports to Europe were five times those to the USA, and 44% higher than those to Japan.

3. This development of production, agricultural as well as industrial, from the oil-exporting countries will find part of its market in Europe. Some of these productions have a clear locational or cost advantage and actually represent an application of the principle of the relative trading advantage. They are competitive with European productions with which they must compete in a non-destructive way.

4. The Arab world is made up of rich and poor countries, the first with a surplus of capital and scarce labour, the second in the opposite situation.

The poor countries are not getting any better in time: many of them are actually getting worse. Many of their young people emigrate to Europe. What stranger sight could there be than a country which thirty years after fighting a bitter war against its colonial overlord finds that its young people consider emigrating to that same former imperial country as the only chance of advancement or even survival!

These countries suffer in various degrees from insufficient capital formation, which goes hand in hand, at least for some of them, with a need to increase the efficiency of the organisation of the economy.

Moving capital from a resource-rich to a resource-poor country would increase productivity all around. It would increase the marginal productivity of the capital of the surplus countries and at the same time reduce their demand for expatriates; and it would employ more of the only resource existing in the poor ones, people. The acceleration of economic growth in the southern Mediterranean shore would reduce the immigration flow in Europe, an element of clear political importance.

These complex elements were arranged in an array of interlocking models used to simulate, on the basis of world economic scenarios, the effects of a regional cooperation policy between the Arab countries and Europe, that is, in a substantial part of what oilmen call the eastern hemisphere.

The model found the answer in economic development, rather than in retrenchment and protectionism. From development Europe can expect the advantages of an international division of labour tailored to the factor costs differentials in the various areas; and the improvements in its well-being and security due to a more stable Mediterranean area. The advantages of such a policy were calculated in terms of extra GDP growth for both producers and consumers (see Appendix: Tables 9.2 and 9.3). The model estimated that the GDP of OPAEC countries would increase by 3.8% from 1980 to 1990 in the neutral simulation, and by 6.2% with cooperative simulations; while OECD GDP would increase by 2.3% in the first simulation and by 3.3% in the second. Oil price would increase by 12.7% per year in the first and by 11.1% per year in the second; total OAPEC exports would increase by 0.3% per year in the first and by 2.5% in the second. To have such a good effect, a good degreee of cooperation would be required in the first instance.

The financial flows from rich to poor Arab countries have been considerable since 1975, and so have population movements. Inter-Arab trade, though negligible at present, offers great possibilities for development. In the second instance, some kind of common development policy should be entertained together by European and Arab countries.

How the situation has changed

As we all know, these words of wisdom remained only words, and attempts to revive the argument got nowhere. There were many reasons for that.

The first was the obvious one, that the interests of the two groups, the exporters and the importers, were seen to be too different; the exporters simply wanting high and rising prices of oil, and the consumers instead wanting a steady, reasonably priced supply.

There was, moreover, a large obstacle of mutual perception. Consumers still remembered the devastating effects of the oil shocks – they never thought that the slump of the late 1970s and early 1980s was due to anything other than that cause; while exporters still saw the rich and strong begrudging the "fair price" to the poor and weak. And the exporters did find it very difficult to accept the market power of the consumers, which had never been included in their own admittedly overoptimistic calculations.

The second was a matter of economic situation.

The oil-rich countries, it was said at that time, are not that rich, and their period of plenty might not last for long. In fact, the decline of the revenues had been eating even the receipts of the best producers; and the economic problems of some of the "poor" Arab countries seemed to be intractable.

The matter of perception was at work here too, which made the exporters quite sceptical.

In the 1970s, when the market was clearly against them, the importers simply tried to adapt to it. Can anybody expect them to adopt a different attitude at any other time? They may just repeat their answer of the 1970s and early 1980s: they will simply react to the market, whatever price it produces.

And we could continue; but we believe that it is more important to present the reasons why these different perspectives are no longer relevant, that they are simply something from the past.

First, the matter of oil prices. Oil prices are not fixed by exporters, but by the futures market, which may be fickle, speculative and uncertain, but it is certainly not run by the producers. Not a single OPEC crude is among the price setters, and the "marker crude" function of the Arabian Light is now a matter of past history. The futures market cannot be interpreted as being "fixed" by the sellers, or, for that matter, by the buyers.

Of course, the new way of fixing prices has created its own problems.

Oil companies do occasionally complain that they have been dispossessed of the power to fix the price of crude by the financial institutions and the brokers which operate on the futures market. They point out that such financial institutions can throw into the futures' game such an amount of resources that no other element is of any importance, and the market "fundamentals" – the level and trend of demand and supply – have become only a basis for speculation and short–term gain. Oil exporters could very well point out that having used the futures market, as it gradually evolved from the spot market, as an important weapon to wrestle the price of oil from their hands, the companies have only themselves to blame for having jumped from the frying pan directly onto the burning coals.

Whatever the position of the oil companies or, at least, of some of them, the consumers have no more reason to be suspicious of the exporters' motivation, and confidence between the two groups of countries should by now feel its positive effect.

Moreover, the exporters have effectively given a pledge of being "long-term suppliers", that is, of being able to follow the increased oil demand of the world's economy by increasing their productive capacity. The resources which were invested in a number of OPEC and non-OPEC countries to increase reserves and to maintain and develop the auxiliary facilities bear witness to this pledge.

So, consumers and exporters should by now have crossed the threshold of trust in each other, and should now be able to deal by consensus with the problems which will inevitably arise in time.

Second, the argument of the economic situation of both the groups of countries. Let us talk first of Europe, with an argument which, I believe, could also be applied to the USA.

The mature European economy needs a breathing space to sustain its growth rate, as well as a strong wave of investment to increase demand for its capital and consumer goods. The failure to broaden the investment base of the economy is perhaps the real reason for sluggish growth in Europe even since the fall in oil prices, and the deep recession ending in the mid-1990s shows this with great clarity.

Europe finds itself at a junction with only two exits: protectionism on one side and market enlargement on the other. The first has always in European history been associated with slow growth: today it would be even riskier than in the past. Among other things Europe should deny free entry in its own market to goods coming from areas from which the vital oil supply comes and which have a high level of import from Europe. A two-way cooperation would fit Europe even more than the Arab producers.

The second way – market enlargement – now offers a great opportunity not only on the southern shores of the Mediterranean, an area which will either develop with Europe's help or continue into an even deeper political unrest; but also to the whole Middle East, where investment would generate more demand for capital goods, which is quite stagnant in Europe, and for industrially produced consumer goods.

In such a framework, investment would be used for the implementation of projects in industry, agriculture and services, organised in such a way as to take advantage of European technology, know-how and marketing; of the capital available in Arab countries as well as in Europe; and of the development opportunities existing in the overpopulated, underdeveloped Mediterranean countries. The project mode, although effective in preventing capital waste, could be implemented in conditions which would allow the control of both domestic inflation – stimulated by the investments themselves – and the pressures on the balance of payments of the developing countries. These two problems, intractable in isolation, would be easier to solve in a cooperative framework.

To go back for a moment to the oil price, let us say that, if such a consensus could be crystallised, the economic relations between oil exporters and importers would be put on a safer basis, and the price of oil would become a part of an agreement, and perhaps not the major one. Cooperation would not depend on any specific price level, or even on the level of the producers' oil income. The surplus coming from oil sales would help to finance a policy of common growth and to achieve a certain level of investment in the less-favoured countries, to be financed by the two partners according to their means and their economic circumstances.

Of course, some surplus countries might object that after 1986 the surplus was somewhat reduced. However, the oil-rich countries have remained surplus ones, needing only the application of normal efficiency principles in the domestic market to make such a surplus reappear.

Finally, the exporters cannot now avoid recognising that the consumers have gained some power over the oil economy and that a *modus vivendi* has to be reached with them. In the same way as the exporters control a large part of the crude oil supply, so the importer countries influence the markets for oil products and can manage demand in the short term by taxation and by stocking and destocking; in the medium and long term by fiscal policies, conservation and diversification. As the experience of the 1970s and 1980s has shown, a lack of understanding, or, worse, disagreement between these two cannot lead to stability, or orderly development in the medium or long term: at best it would produce price volatility and a seesaw of shocks and collapses. The interests of buyers and sellers are not identical, but do coincide to a certain degree. There is no contract without mutual interest; seller and buyer can exist as such only if both agree to enter into and remain in such a relationship.

The main difference between the two groups of countries stems from what we pointed out before, that oil is not as important for the importer as it is for the exporter, being for the former merely a cost item in the production of goods and services, and an expense item for the household. The importers' strategy will be centred not around oil, but rather around the state of their economies. There is therefore no way of reaching, or even of discussing, a consensus without putting the oil business in the wider perspective of the world economy.

The exporters never actually refused the idea of moving from an oil policy to a wider economic policy, which is also the consequence of their attempts to diversify their economies, stemming from the refusal to let a single product, crude oil, remain forever at the centre of their economic life. However important, oil must gradually become for them also just one of the elements of a wider picture, which has as its main concern the rate of growth of the domestic and world economy.

A structural element of the oil industry: the role of the State

The present set-up of the oil industry has a number of important structural elements.

The first is that a large part of the oil production – the part of the exporters' countries – is obtained not by commercial companies, but by companies owned by States and led by Ministers who take decisions primarily according to the institutional system of the country rather than on market results. These "companies" are technical operators without strategic responsibilities and sometimes without a full accounting system for costs and revenues. The money from the sale of oil flows back to the State (to the Central Bank, or the Treasury) to pay for the "fiscal take" (which is by definition equal to the difference between actual costs and prices) and it is used to finance the State budget. As the exporting countries obtain from these sales the largest component of their GDP, it is quite understandable that oil matters should go for decisions to the highest echelons of the State.

The price of oil represents for these countries what the fiscal pressure on the citizens' income is for the other countries in the world. Many oil exporters have no tradition of tax-paying, and in many even the basic elements for a personal income tax, like a fiscal file for every citizen, are either not present or very uncertain and unreliable. Sometimes even the very distinction between the State's Treasury and that of the ruling family is recent and still somewhat uncertain.

So, the oil strategy becomes an essential part of the economic policy of a country, which must include many elements which go well beyond the oil industry, and relate to the strength of the State, the welfare of the citizens, etc..

The State has an important position in consuming countries, too, and not only *vis-à-vis* the oil supply, and the consequent "oil bill". In general, governments are still seen as being responsible for the welfare of the citizens, and having overriding interest in the growth of their national economy.

As far as energy is concerned, industrial or household consumers are too scattered and differentiated to have any power on the energy market. It is up to their government to represent and to defend the interests of the nation's consumers.

These countries are pooled in the International Energy Agency, which has set up a system for prevention of shortages and for resource allocation in case shortages occur. Only government officials can be involved day-to-day in operations like that, although they might require some advice from people from the industry (which is just the way the Agency works).

The fact that many consumer's states do tax energy in general, and gasoline in particular, does not, in their view, change the issue, which is seen to be related to the expenditure to acquire the supply, which shapes the balance of payments of these countries, and eventually the position of national currencies.

In short, the governments of both consumers' and exporters' countries seem to be in a similar positions *vis-à-vis* their citizens and the general level of economic life in the country.

It should follow that some kind of cooperation or, at least, of reciprocal communication between the two should be relatively easy to implement.

A modest proposal

Both groups of countries should define their economic objectives in terms of steady economic growth world-wide, with an accelerated development in the poorer areas. So, they should welcome the growth of world trade, the main engine of economic progress, and the industrialisation of poor areas, useful to reduce overcrowding in the old countries, and to cash in the opportunities offered by investments in areas with a very high marginal productivity of capital. Such an economic growth scenario needs an abundant oil supply at a reasonable price; and such a supply could very well become one of the instruments for the expected growth.

If consumers and exporters have the same objective, they could very well pursue it together, now that the contentious issue of oil prices has fallen into the hands of a market which is scarcely prone to manipulation, and can therefore be considered "impartial". They are now tied to each other by the fact that the one sells what the other wants to buy; they could build on this tie by activating together a policy to use the opportunities created by the flow of oil to accelerate world economic development. Exporters and importers are interdependent not only because they are made so by the distribution of oil reserves (which cannot be changed) and energy consumption (which changes every day), but also because they share the opportunity offered by this peculiar position.

Interdependence is not a new concept: it has, up to now, been a powerful idea, and a spur to action.

Appendix: The advantages of interdependence

The quantitative simulations of the Interdependence System of Models (ISM) show the positive effects of cooperation policies both on oil matters and on general economic growth. The first run, done in 1980, was basically a comparison between two scenarios: the Neutral System Interaction (NSI), representing more or less a continuation for the decade 1980–1990 of what had been happening from 1974 onwards, including fine-tuning of supply and consumers' reactions, and another, called Cooperative System Interaction (CSI), more or less inspired by the Taif formula. The first simulation was done on the oil policy, using prices and scenarios compatible with the atmosphere of the late 1970s; it may appear out of tune with reality. But whatever the absolute figures, the difference between the two simulations is nevertheless quite relevant.

The results of the first simulation, presented below, show that between 1980 and 1985 a slightly slower increase in the crude oil price would have caused OAPEC production to increase by 1.6 mb/d, against a loss of 2.5 mb/d in the "Neutral" simulation.

Oil income would have been $240 million higher; and the positive difference would have been much higher in 1990 (Table 9.1).

When these results were presented in Rome, in April 1981, the second oil shock had made sure that any cooperative solution remained on paper.

The second simulation was aimed at GDP and economic development. Although it is in part subject to the same criticism as the first one, it may be useful to present it (Table 9.2).

Table 9.1
FIRST SIMULATION: THE OIL POLICY
Oil production, prices and income of OAPEC countries

	Neutral System: Interaction (NSI)			*Cooperative Strategies: Interaction (CSI)*	
	1980	*1985*	*1985*	*1985*	*1990*
Oil exports (mb/d)	19.8	17.3	18.3	21.4	24.2
Average oil price in current $/b	31.2	65.0	103.5	55.6	89.6
Oil income current ($billion)	2,220	4,050	6,118	4,290	7,820

Table 9.2
SECOND SIMULATION: MACROECONOMIC DATA:
average growth rates from 1980 to 1990

	NSI	*CSI*
OAPEC GDP	3.8	6.2
OAPEC non-oil GDP	6.5	9.0
OECD GDP	2.3	3.3
Oil price	12.7	11.1
Relative oil price	4.4	3.0
OAPEC exports	0.3	2.5
Of which:		
Oil	–0.7	2.0
World trade	3.2	4.9

Table 9.3
COMPARISON BETWEEN TREND AND EURO-ARAB SIMULATIONS OF THE ISM
REAL AVERAGE ANNUAL PERCENTAGE CHANGE 1985–1995

		Trend	*Euro-Arab*
Arab countries			
Arab 1 (Algeria, Egypt,	Investment	3.5	6.5
Iran, Syria)	GDP	4.4	7.6
Arab 2 (Saudi Arabia,			
Bahrain, Kuwait, Libya,	Investment	3.0	7.0
Qatar, UAE)	GDP	3.7	5.1
Arab 3 (Tunisia,			
Morocco, Yemen AR,	Investment	4.0	6.0
Oman, Jordan)	GDP	3.4	5.0
European countries (GDP only)			
UK		1.6	2.0
France		2.5	3.0
Germany		2.1	2.2
Italy		3.0	3.4
Spain		2.5	3.5

Producers and consumers would both have developed faster in the cooperative solution than in the other, the differences being serious for both, but being much larger for the producers than for the consumers.

The second run of the system of Models, operated by a mixed Arab–European team, simulated among others a scenario called "Euro-Arab Cooperation with Constant Oil Prices" which entailed a set of agreements between Europe and the Arab countries belonging to OAPEC. This scenario was then confronted with the reference one, called "TREND".

In the Europe–Arab scenario conditions were pushed to the extreme, in order to demonstrate that the advantages of cooperation did not disappear for Europe even in an exaggerated scenario, with conditions which could possibly be considered tipped in favour of the Arab countries. In this scenario, the price of oil remains constant, i.e. reaches 50 current dollars per barrel in 1995 starting from 27 in 1985; but the final price to European consumers actually decreases in national currencies by about 5%, because of a reduction of taxes on petroleum products. An increase in agricultural exports from Arab countries is also envisaged, and a preference is given to Arab oil and other products. The Arab oil producers adopt corresponding policies on crude oil payments (for example quoting them in European currencies) and give preferential treatment to Europeans in selected areas of investment in agriculture, industry, oil exploration and refining. These policies have positive effects which permit a policy of fiscal expansion in Europe, while cooperation among Arab countries increases the investments in that area. The 1985–1995 annual growth rates of GDP are higher than those of the reference scenario: in Europe up by a maximum of a whole percentage point for Spain, and by about half a point for the UK, France and Italy, West Germany being the least affected. The growth rates of the Arab countries are of course much higher, up by three percentage points for the poorer countries and between one and two for the others (Table 9.3).

Of course, the scenarios simulated with the Interdependence System of Models may be taken simply as an example: others could be tested, chosen with an eye to political feasibility, which was not paramount in the choice of the interdependence experts.

The results, however, are clear: even an increase in the current price of oil, a pretty extreme condition in 1985, would not cancel the advantages of a policy of cooperation between Arab oil producers and European oil consumers.

Annex 1

DEFINITIONS AND CLASSIFICATIONS

Petroleum Includes crude oil, natural hydrocarbon gas and the condensate content of natural gas, wherever these occur under natural conditions.

Field Refers to a single accumulation (in which case it is synonymous with "pool" or "reservoir"), or a set of closely related accumulations of petroleum, containing oil and gas in substantial quantities, which in most cases means in commercial quantities. . It is a producing area containing in the sub-surface: (1) a single pool uninterrupted by permeability barriers, (2) multiple pools trapped by a common geological feature, or (3) laterally distinct multiple pools within a common formation and trapped by the same type of geological feature.

Petroleum in place The original amount of petroleum present.

Reserves The recoverable proportion of the petroleum in place remaining to be produced/discovered, less production to date.

Proven Reserves The amount of discovered petroleum that can reasonably be expected to be produced under present and expected local economic conditions with existing available technology.

Probable Reserves The amount of discovered petroleum that can reasonably be expected to be produced from the field, less proven reserves. This includes extensions to the limits of the field, incremental production attributable to as yet uninstalled improved recovery projects, etc.

Ultimate Reserves Past production plus reserves, proven and to be found (undiscovered potential).

Province A basic regional geological unit for petroleum exploration and production.

"Major" oil producing province A province with a known oil recovery of ten billion barrels or more or one which, with further discoveries, development and implementation, is likely to reach ten billion barrels or more.

M. Colitti and C. Simeoni, Perspectives of Oil and Gas: The Road to Interdependence, 137–139.

Undiscovered A petroleum accumulation whose existence has not been demonstrated by the drill, but is merely postulated – for example, possible reserves in an undrilled trap.

New field wildcat (NFW) A hole drilled on a geological trap where petroleum has not yet been discovered. The wildcat is a well drilled far from any other – as far as a wildcat could jump – in order to find a new field. It is normally the conclusion of the process of indirect exploration.

Dry Hole Any hole in which petroleum has not been discovered.

Discovery Well Any wildcat, deeper pool test or shallower pool test which discovers a new pool of petroleum.

Date of Discovery The date on which the discovery well of the pool is completed as a producer of petroleum.

Classification of Fields Fields are classified by size, namely total production to date and proven reserves that can be extracted at the time the estimate is made. The definition "extractable reserves under present conditions" is generally understood to include secondary oil recovery, provided that recovery operations have been computed and all the necessary arrangements have been made.

The size classification of the fields originated some 50 years ago in the USA. The breakdown into classes was based on the size of the US fields, which are mostly in the small to medium range; so a "major" is a field with reserves exceeding 100 million barrels. However, in recent years a classification has been evolved based on the worldwide situation, not merely that in the USA. This involves a breakdown into classes on a much more analytical basis where the major fields are concerned, with distribution and weight of the highest classes being taken into account.

The two forms of classification are given below. One does not exclude the other: they are mutually complementary. The new classification has been shown to be exceptionally significant.

US classification

Major fields:	more than 100 million barrel recoverable oil
Class A fields:	at least 50 million barrel recoverable oil
Class B fields:	between 25 and 50 million barrel recoverable oil
Class C fields:	between 10 and 25 million barrel recoverable oil
Class D fields:	between 1 and 10 million barrel recoverable oil
Class E fields:	less than 1 million barrels of recoverable oil
Class F fields:	abandoned

World classification

Super giant fields:	more than 5,000 million barrels of recoverable oil
Large giant fields:	between 2,000 and 5,000 million barrels of recoverable oil
Medium giant fields:	between 1,000 and 2,000 million barrels of recoverable oil
Small giant fields:	between 500 and 1,000 million barrels of recoverable oil
Large no-giant fields:	between 100 and 500 million barrels of recoverable oil
Medium-size fields:	between 25 and 100 million barrels of recoverable oil
Small-size fields:	between 10 and 25 million barrels of recoverable oil
Very small fields:	less than 10 million barrels of recoverable oil

Annex 2

CHRONOLOGY OF THE 97 MEETINGS OF THE OPEC CONFERENCES[1]

from 1960 to 1994 (excerpts on decisions settled on oil productions and prices)

The first move towards the establishment of the Organization of the Petroleum Exporting Countries – OPEC – took place in 1949 when Venezuela approached Iran, Iraq, Kuwait and Saudi Arabia and suggested that they exchange views and explore avenues for regular and closer communications between them.

The First Arab Petroleum Congress, held in Cairo, adopted a resolution calling on oil companies to consult with the governments of the producing countries before unilaterally taking any decision on oil prices, and set up the general agreement on the establishment of an "Oil Consultation Commission".

On September 10–14, 1960, the conference held in Baghdad, attended by representatives of the governments of Iran, Iraq, Kuwait, Saudi Arabia and Venezuela, established OPEC as a permanent, intergovernmental organization.

Structure of OPEC

I **The Conference** is the supreme authority of OPEC, and consists of delegations headed by the Ministers of Oil, Mines and Energy of Member Countries.

The Conference, which meets at least twice a year and operates on the principle of unanimity, is responsible for the formulation of the policy of the Organization and the determination of the appropriate ways and means of its implementation.

II **The Board of Governors** directs the management of the Organization; implements resolutions of the Conference; draws up the Organization's annual budget and submits it to the Conference for approval.

III **The Secretariat** carries out the executive functions of the Organization, in accordance with the provisions of the Statute and the direction of the Board.

IV **The Secretary General** is the legally authorized representative of the Organization and Chief Executive of the Secretariat in which capacity he administers the affairs of the Organization in accordance with the directions of the Board of Governors.

---

[1] *Source*: *OPEC Bulletin*, various issues, Vienna, Austria.

M. Colitti and C. Simeoni, Perspectives of Oil and Gas: The Road to Interdependence, 141–158.

V **OPEC's Principal Aims** are the co-ordination and unification of petroleum policies of Member Countries and the determination of the best means for safeguarding their interests, individually and collectively. The Organization also seeks to devise ways and means of ensuring the stabilization of prices in international oil markets with a view to eliminating harmful and unnecessary fluctuations, due regard being given at all times to the interests of the producing nations and to the necessity of securing a steady income for them; an efficient, economic and regular supply of petroleum to consuming nations; and a fair return on their capital to those investing in the petroleum industry.

In June 1968, a "Declaratory Statement of Petroleum Policy in Member Countries" was adopted and in March 1975, the 'Solemn Declaration' added new policy guidelines in the light of the changing pattern of the relationship between producer and consumer nations.

Membership: any country with a substantial net export of crude petroleum, may become a Full Member of the Organization, if accepted by a majority of three-fourths of Full Members, including the concurring votes of all Founder Members. The Statute further distinguishes between three categories of Membership: Founder Member, Full Member and Associate Member.

Founder Members of the Organization are those countries which were represented at the First Conference, held in Baghdad in 1960, and which signed the original agreement establishing OPEC.

Full Members are the Founder Members, plus those countries whose applications for Membership have been accepted by the Conference. An Associate Member is a country which does not qualify for full membership, but which is nevertheless admitted under such special conditions.

The Organization now has 12 Member Countries. They are listed below with their year of accession.

Founder members	**Full members**	
Islamic Republic of Iran	Qatar	1961
Iraq	Indonesia	1962
Kuwait	Socialist People's Libyan Arab Jamahiriya	1962
Saudi Arabia	United Arab Emirates	1967
Venezuela	Algeria	1969
	Nigeria	1971
	Gabon (Associate Member 1973–75)	1975

The OPEC Fund for International Development is dedicated to promoting co-operation between OPEC and other developing countries. It pursues this aim primarily by helping provide the financial resources those countries need to realize their economic and social development goals.

All non-OPEC developing countries are, in principle, eligible for Fund assistance.

OPEC and other international organizations

OPEC came into being as a result of a joint action undertaken by several countries which are members of the United Nations. The treaty establishing OPEC was registered at the United Nations Secretariat on November 6, 1962.

In 1965 OPEC also established relations with the Economic and Social Council.

The OPEC Secretariat has become a regular participant in meetings of various organs, bodies and specialized agencies of the United Nations, particularly the United Nations Conference on Trade and Development, the World Bank, the International Monetary Fund and the international fund for agricultural development.

Secretaries General of OPEC (1961–1995)

Name	*Member country*	*Term of office*
Dr Fuad Rouhani	Iran	21.1.1961 – 30.4.1964
AbdulRahm Al Bazzaz	Iraq	1.5.1964 – 30.4.1965
Ashraf Lutfi	Kuwait	1.5.1965 – 31.12.1966
Mohammad S. Joukhdar	Saudi Arabia	1.1.1967 – 31.12.1967
Francisco R. Parra	Venezuela	1.1.1968 – 31.12.1968
Dı Elrich Sallgcr	Indonesia	1.1.1969 – 31.12.1969
Omar El-Badri	Libya	1.1.1970 – 31.12.1970
Dr Nadim Pachachi	United Arab Emirates	1.1.1971 – 31.12.1972
Dr Abderrahman Khene	Algeria	1.1.1973 – 31.12.1974
Chief M.O. Feyide	Nigeria	1.1.1975 – 31.12.1976
Ali M. Jaidah	Qatar	1.1.1977 – 31.12.1978
Rene G.Ortiz	Ecuador	1.1.1979 – 30.6.1981
Dr Marc S.Nan Nguema	Gabon	1.7.1981 – 30.6.1983
Dr Fadhil j. Al- Chalabi	Iraq	1.7.1983 – 30.6.1988
Dr Subroto	Indonesia	1.7.1988 – 30.6.1994
HE Abdalla El Badri	Libya	1.9.1994 – 31.12.1994
Dr Rilwanu Lukman	Nigeria	1.1.1995

Chronology

1st, Baghdad, September 10–14, 1960: "The first OPEC Conference was held in Baghdad, a statement of Policy and Objectives was adopted and Membership requirements established".

5th, Riyadh, December 24–31, 1963: "... Establish a three-member OPEC negotiation committee (Iran, Iraq and Saudi Arabia) with the oil companies on royalty payments and marketing expenses; "

16th, Vienna, June 24–25, 1968: "... Adopted the 'Declaratory Statement of Petroleum Policy in Member Countries' emphasizing, inter alia, that Member Governments should endeavour to explore for and develop their hydrocarbon resources directly; acquire reasonable participation in the ownership of the concession-holding companies; participate in choosing the acreage to be relinquished; and base the assessment of the companies' income, taxes and any payment to the State on of the principal sources of revenue and foreign exchange earnings for Member Countries and the proper exploitation of these resources would determine the conditions of their economic development, both in the present and in the future. It also emphasized the inalienable right of all countries to exercise permanent sovereignty over their natural resources in the interest of their national development, the exploitation of which natural resources should be aimed at securing the greatest possible benefit for Member Countries. This aim could better be achieved if Member Countries themselves were in a position to undertake, directly, the exploitation of these resources so that they would be able to exercise their freedom of choice to utilize them under the most favourable conditions".

17th, Baghdad, November 9–10, 1968: "... Recommended that Member Countries should not grant any new oil rights to companies whose home countries pursued policies tending to artificially depress petroleum prices on international markets; and recommended that Member Countries should, subject to differences in gravity quality and geographical location, seek to ensure that posted or tax reference prices of their petroleum exports were consistent."

21st, Caracas, December 9–12, 1970: "... To establish 55% as the minimum rate of taxation on the net income of oil companies; to eliminate disparities in posted or tax reference prices of crude oils on the basis of the highest posted price applicable; to eliminate completely the allowances granted to oil companies as from the beginning of 1971..."

Tehran, February 14, 1971: "...The Tehran Agreement was signed between six Member Countries in the Gulf and 23 international oil companies whereby the posted price of the Arabian Light was adjusted to $2.18/barrel. The agreement included a flat across-the-board increase of 33 cent/b in the posted prices, a flat increase of 5 cents anually in lieu of claims relating to adjustments in product prices, an increase of 2.5 per cent annually for inflation, full expensing of royalties, elimination of the marketing allowances, an increase in the tax rate to 55 per cent, and the realignment of all posted prices in the Gulf, using the price of Arabian Light as a reference, taking into account a freight differential based on a rate of Worldscale 72."

Geneva, January 20, 1972: "The Geneva I Agreement was reached between OPEC Countries and the oil companies, supplemental to the 1971 Tehran Agreement, whereby oil revenues of Member Countries were increased by the provision of an immediate price adjustment in the Arabian Light posted prices of 8.49 per cent. The settlement also included provisions for further adjustments in oil revenues between 1972 and 1975 based on an index measuring changes between the exchange rates of the dollar and nine other key currencies."

Geneva, June 1, 1973: "... Geneva II Agreement was reached with the international oil companies on adjustments to posted prices to take account of the world currency situation; it included an expanded 'basket' of major currencies, from 9 to 11, with a new mechanism for a monthly measurement of change..."

Kuwait, October 16, 1973: "The Gulf Members announced an immediate increase in posted prices to $5.12 per barrel. It was the first move by Member Countries to exercise their sovereign right to determine the price of their natural resources. From this point on, OPEC set prices unilaterally for its oil."

Tehran, December 22–23, 1973: "... Decided to set the Government take at $7 a barrel for Marker Crude and, therefore, taking into account the Geneva II Agreement, the posted price for Marker Crude would become $11.651/barrel as of 1 January, 1974..."

37th, Geneva, January 7–8, 1974: "...The Gulf Member Countries decided to abandon the ratio of 1.4 to 1.0 between posted prices and market prices, in response to changes in the petroleum market conditions."

40th, Quito, June 15–17, 1974: "... Decided to increase the rate of royalty by two per cent or alternatively to increase the Government take by an equivalent amount. Bolivia, Colombia. Congo, Peru, and Trinidad and Tobago attended as observers."

41st (Extraordinary), Vienna, September 12–13,1974: "Decided to freeze posted prices for the fourth quarter of 1974, but to increase the total government take by 3.5 per cent, effective October 1, 1974 in order to compensate for inflation in the industrialized countries. The weighted average government take for Marker Crude would thus increase from $9.41 to $9.74 per barrel.

The weighted average government take would become the average between the government take on equity oil and the government take on participation oil. As a result royalty payments would increase to 16.67 per cent, the remainder representing an increase in the applicable tax rate of each Member Country..."

Abu Dhabi, November 10, 1974: "Three OPEC Gulf States, Saudi Arabia, United Arab Emirates and Qatar, decided to raise income tax and royalty rates on the equity crude lifted by the major oil companies to 85 per cent and 20 per cent respectively."

42nd, Vienna, December 12–13, 1974: "Decided to adopt a new pricing system based on the decision taken by the three Gulf States on November 10. As a result, the average government take was set at $10.12 per barrel for Marker Crude, effective from January 1, 1975 through to September 30, 1975..."

Algeria, March 4–6,1975: "The Conference of Sovereigns and Heads of State of OPEC Member Countries adopted the 'Solemn Declaration' which emphasized, inter alia, mutual respect for the sovereignty and equality of all member nations of the international community in accordance with the UN Charter, and reaffirmed the sovereignty and the inalienable right of all countries to the ownership, exploitation and pricing of their natural resources. They noted that the disequilibrium which affected the international economic situation was aggravated by widespread inflation, a general slowdown of economic growth and instability of the world monetary system due to the absence of monetary discipline and restraint. Causes of such anomalies lay in the general tendency of developed countries to consume excessively and to waste scarce resources. The unilateral decision by OPEC Member Countries to adjust their oil prices had contributed but insignificantly to the high rates of inflation which had been generated within the economies of the developed countries. The Sovereigns and Heads of State declared the readiness of OPEC Member Countries, within the framework of their solidarity, to take immediate and effective measures to counteract threats against their sovereignty, and reaffirmed the solidarity which united their countries with the other developing countries in the struggle to overcome under development. They stated that OPEC Member Countries were prepared to contribute to the realization of the UN special international programme and to extend additional special credits, loans and grants for the development of developing countries. They reaffirmed their willingness to co-operate with those other developing countries that were exporters of raw materials and other basic commodities, in their efforts to obtain an equitable and remunerative price level for their exports. As regards the supply of petroleum, they gave the assurance that sufficient supplies would be made available to meet the essential requirements of the economies of developed countries.

With respect to petroleum prices, they pointed out that high rates of inflation and currency depreciation had wiped out most of the real value of OPEC oil price adjustments and that posted prices were markedly lower than prices of conditions of availability, utilization and costs of alternative sources of energy. The Declaration expressed the agreement of the OPEC Sovereigns and Heads of State to the holding of an international conference bringing together developed and developing countries..."

45th (Extraordinary), Vienna, September 24–27, 1975: "Decided to increase the price of Arabian Light, the Marker Crude, by only ten per cent, bringing it to $11.51 per barrel from October 1, 1975, and hold it at that level until June 30, 1976..."

48th, Doha, December 15–17, 1976: "Decided, with the exception of Saudi Arabia and the United Arab Emirates, to increase the price of Marker Crude by ten per cent from $11.51 per barrel to $12.7 per barrel as of January 1, 1977, and by a further five per cent to $13.3 as of July 1977.

The prices of all other crudes were to be adjusted in accordance with the new prices of Marker Crude.

Saudi Arabia and the UAE decided to raise their prices by five per cent only..."

49th, Stockholm, July 12–13, 1977: "Algeria, Ecuador, Gabon, Indonesia, Iran, Iraq, Kuwait, the Socialist Peoples Libyan Arab Jamahiriya, Nigeria, Qatar and Venezuela formally agreed to forgo the application of an additional five per cent increase in oil prices.

Saudi Arabia and the United Arab Emirates adjusted their prices upwards, which brought an end to the two-tier pricing system..."

52nd, Abu Dhabi, December 16–17, 1979: "... Decided to adjust the price of oil by a ten per cent average over the whole of 1979, broken down into the following quarterly adjustments on a cumulative basis: five per cent on January 1, 1979; 3.809 per cent on April 1, 1979; 2.294 per cent on July 11 1979; and 2.691 per cent on October 1, 1979.

The application of these adjustments to Marker Crude would yield the following prices: $13.335 as from January 1, 1979; $13.843 as from April 1, 1979; $14.161 as from July 1, 1979; and $14.542 as from October 1, 1979..."

53rd (Extraordinary), Geneva, March 26–27, 1979: "... Decided to bring forward the price adjustment of the fourth quarter of 1979 and apply it as of April 1, 1979. This adjustment set the Marker Crude at $14.546..."

54th, Geneva, June 26–28, 1979: "... Decided to adjust the Marker Crude price to $18 per barrel, to allow Member Countries to add to the prices of their crude a maximum market premium of $2 per barrel over and above their normal differential, if and when such a market premium was necessitated by market conditions; and that the maximum prices that could be charged by Member Countries were not to exceed $23.50."

56th (Extraordinary), Thaif, May 7–8, 1980: "... Recommended that the price of natural gas should be in line with the price of oil on a BTU basis..."

57th, Algiers, June 9–11, 1980: "... Decided to set the price level for a Marker Crude at a ceiling of $32 per barrel; that the value differentials which would be added over and above this ceiling for the Marker Crude, to take account of quality and geographical location, should not exceed $5 per barrel; and that this price structure should apply as of July 1, 1980. The objective of the new price structure was to achieve an equilibrium between supply and demand to avoid further stockpiling, which was considered harmful to both producers and consumers. The Conference reiterated OPEC Member Countries' determination to set gas prices in line with those of crude oil in order to achieve a coherent marketing policy for their hydrocarbons..."

58th, Vienna, September 17, 1980: "... Decided to fix the price of the Marker Crude (Arabian Light) at the level of $30 per barrel and to freeze the other official prices of OPEC Member Countries' crudes at that level..."

59th, Bali, December 15–16, 1980: "...Decided to fix the official price of the Marker Crude at the level of $32 per barrel; that prices of OPEC crudes may be set on the basis of an oil price ceiling for a deemed Marker Crude of up to $36/ barrel; and to set the maximum price for OPEC crudes at $41/barrel."

60th, Geneva, May 25–26, 1981: "... Decided to maintain the deemed Marker Crude price at a ceiling of $36 per barrel, with a maximum OPEC price of $41/b, until the end of the year. The majority of Member Countries decided to cut production by a minimum of ten per cent with effect from June 1, 1981."

61st (Extraordinay), Geneva, October 29, 1981: "Recognizing the need to adopt a unified pricing system for OPEC crudes with a view to creating the right conditions for stability in the market, resolved to set the official price of the

Marker Crude at $34 a barrel, effective not later than November 1, 1981, and to abide by that price until the end of 1982. The Conference also agreed to a set of value differentials for the pricing of all other OPEC crudes in accordance with their respective qualities and geographical locations."

63rd (Extraordinary), Vienna, March 19–20, 1982: "... Decided to reconfirm the price of the Marker Crude, Arabian Light, 34° API, ex Ras Tanura, at $34/b as had been decided by the 61st Conference, and to take the necessary measures for stabilizing the market. For this purpose it was decided that, as of April 1, 1982, total OPEC production would have a ceiling of 18 milion b/d. It was further decided that the price differential for light and extra light crudes in relation to the Marker Crude would be set at the same levels as in 1978..."

66th, Vienna, March 19–20, 1982: "... Decided that OPEC production for the year 1983 should not exceed 18.5 million b/d..."

67th (Extraordinary), London, March 14, 1983: "Resolved to:

1. set the official price of the Marker Crude, Arabian Light, 34° API, ex-Ras Tanura, at $29/b.

2. maintain the existing differentials among the various OPEC crudes at the same level as agreed upon at the 63rd Conference with the temporary exception that the differentials for the Nigerian crudes should be $1 over the price of the Marker Crude.

3. establish a ceiling for total OPEC production of 17.5 million b/d, within which individual Member Country quotas were allocated. This ceiling was to be observed as an average for the remaining part of 1983. No quota was allocated to the Kingdom of Saudi Arabia which would act as a swing producer to supply the balancing quantities to meet market requirements."

71st (Extraordinary), Geneva, October 29–31, 1984: "... Determined to defend the price structure of its Marker Crude (Arabian Light, 34° API, ex-Ras Tanura) at the level of $29 per barrel and to consolidate market stability. The Conference decided to cut, as from November 1, 1984, and on a temporary basis, the global production ceiling of OPEC from 17.5 million b/d to 16 million b/d."

72nd, Geneva, December 19–21 and 27–29, 1984: "Welcomed representatives of Brunei Darussalam, Egypt, Malaysia and Mexico who attended the meeting as observers. (...)

Expressed its determination to enforce the implementation of its decisions on production and prices as a means to defended its price structure, based on the price of $29/barrel for the Marker Crude, Arabian Light.

Agreed, with the exception of Algeria and Nigeria, on the following adjustment of price differentials to be applicable temporarily and reviewed not later than end of January 1985:

1. To keep the Marker Crude price at its current level of $29/barrel.
2. To adjust the prices of heavy crudes by a premium of $0.50/barrel.
3. To adjust the prices of medium crudes by a premium of $0.25/barrel.
4. To adjust the prices of extra-light crudes by $0.25/barrel..."

73rd (Extraordinary), Geneva, January 28–30, 1985: Agreed that the maximum price differential between heavy crude oil produced in the Gulf (Arabian Heavy) and light crudes produced in Africa should be in the order of $2.40 per barrel and decided, therefore, to fix the price of their own crudes with effect from February 1, 1995. However, Algeria, the Islamic Republic of Iran and the Socialist Peoples Libyan Arab Jamahiriya dissociated themselves from this decision. Gabon abstained.

	$/barrel
Arab Heavy	26.50
Arab Medium	27.40
Arab Light	28.00
Murban	28.15
Minas	28.53
Bonny Light	28.65

74th, Geneva, July 22–25, 1985: "... Decided by majority to adjust downward the price of medium crude of 31° API by $0.20 per barrel and Arabian Heavy of 27° API by $0.50 per barrel. Algeria, the Islamic Republic of Iran and the Socialist Peoples Libyan Arab Jamahiriya disagreed with the price reduction and dissociated themselves from this decision..."

75th, Vienna, October 3–4, 1985: "... Agreed to maintain the OPEC production ceiling of 16 million b/d, however, the majority agreed that the redistribution of that ceiling among Member Countries be postponed to a later date..."

77th (Extraordinary), Geneva, March 16–24 (reconvened) April 15–21 1986: "Ten Member Countries agreed that the average OPEC production level for 1986 should be realistically set at 16.7 million barrels per day and that, during the third and fourth quarters of the year, production should be set at 16.3 mb/d and 17.3 mb/d, respectively. They agreed that OPEC Countries would gradually adjust their current production in order to reach the above-mentioned levels. Three Member Countries, namely, Algeria, the Islamic Republic of Iran and the Socialist Peoples Libyan Arab Jamahiriya agreed that the same objective could only be achieved by setting OPEC production for the second, third and fourth quarters of 1986 at 14 mb/d, 14.5 mb/d and 16.8 mb/d, respectively.

Agreed to support the oil price structure, with a view to eventually attaining the restoration of the OPEC reference price to its official level of $28/b."

78th, Geneva, July 28 – August 5, 1986: "... Decided that OPEC's production for the months of September and October 1986 should be fixed on the basis of the ceiling of 16 million barrels per day, as decided in October 1984, distributed among Member Countries with the same national quotas, with the exception of Iraq..."

79th (Extraordinary), Geneva, October 6–22, 1986: "... Decided that the interim and temporary action on Member Countries' production limitation would be extended, with some minor adjustments, through the months of November and December 1986. Consequently, Member Countries' quotas for this new interim period should be:

	Nov 86	Dec 86
	(million b/d)	
Algeria	0.669	0.669
Ecuador	0.221	0.221
Gabon	0.160	0.160
Indonesia	1.193	1.193
IR Iran	2.317	2.317
Kuwait	0.921	0.999
SP Libyan AJ	0.999	0.999
Nigeria	1.304	1.304
Qatar	0.300	0.300
Saudi Arabia	4.353	4.353
UAE	0.950	0.950
Venezuela	1.574	1.574
Total	14.961	15.039

The Conference also discussed a proposal by Saudi Arabia to return to a fixed price structure for OPEC crude oil and decided to return, as soon as possible, to the application of a fixed price system..."

80th, Geneva, December 11–20, 1986: "... Decided to return promptly to a fixed pricing system at a level of $18 per barrel for OPEC's reference price. This price, which was based on a basket of seven crudes, would come into force as of January 1, 1987, with the understanding that a one-month phase-out period would be permitted on current contracts. The Conference went on to discuss the necessary mechanism to defend the OPEC reference price and concluded that in view of the existence of excesssive oil supplies in the market, it was necessary to undertake cuts in OPEC's production level in order to support that price. An agreement was reached that OPEC's total production for the first and second quarters of 1987 should not exceed 15.8 million barrels per day, representing cuts of 7.6 per cent and 7.1 per cent, respectively, from the perceived demand for OPEC oil during these two quarters."

81st, Vienna, June 25–27, 1987: "Expressed satisfaction with the positive results of the Agreement signed in December 1986, under which OPEC re-established the fixed price system at a level of $18 per barrel, that, in order to secure continued firm prices in the market, OPEC production during the fourth quarter of 1987 would be equal to that of the third quarter, i.e. 16.6 million barrels per day, distributed into the same national production levels as had been defined by the above-mentioned Agreement for that quarter..."

82nd, Vienna, December 9–14, 1987: "... The Conference noted the efforts made by the Organization to achieve a fair degree of price stability throughout 1987 and expressed the need for further efforts to be made by Member Countries on production and price performance, with a view to further strengthening market stability. For this purpose, an Agreement was reached that the production level of all Member Countries, with the exception of Iraq, would be 15.06 million barrels per day during the first half of 1988, distributed into national production levels in the same manner as laid down in the December 1986 Agreement for the third quarter of 1987."

Vienna, April 26–27, 1988 – OPEC/Non-OPEC Meeting. "At the invitation of OPEC, seven independent oil producing countries met with a Committee representing six OPEC Member Countries. The independent oil producers present were: Angola, China, Colombia, Egypt, Malaysia, Mexico and Oman. The OPEC Member Countries present were: Algeria, Indonesia, Kuwait, Nigeria, Saudi Arabia and Venezuela.

In a joint communiqué issued at the close of the Meeting, the participants noted that the Meeting constituted the beginning of a new stage of co-operation between a large number of oil producing countries and expressed the hope that other countries would join in the effort. The underlying assumption shared by all participants was that a stable oil market was the responsibility of all producers and, consequently, all producers should share the burden.

The independent producers presented, individually, proposals for bringing prices back to a range around OPEC's objective of \$18/b. In order to achieve this target, the producers acknowledged, production restraint would be necessary, together with the avoidance of destructive commercial practices, like netbacks and other discounting mechanisms. Angola, China, Egypt, Malaysia, Mexico and Oman proposed to cut back their respective export volumes by five per cent during the crucial remainder of the second quarter, i.e. May and June. The basis for such a cut would be the average exports of the past six months. Implementation of such measures would be contingent upon a similar proportional response by OPEC.

The independent producers acknowledged that the OPEC Committee would have to report this proposal to a Consultative Meeting of the OPEC Conference before a response could be given. All participants agreed that the Meeting should be followed by continuing contacts for the purpose of reviewing agreements arrived at and planning future avenues of co-operation. All participants were of the view that the common concern at the present stage must be to establish the long-term nature of the co-operation effort, to guarantee its credibility and to widen its scope through the participation of new countries."

84th, Vienna, November 21–28, 1988: "... Decided to take steps to strengthen the oil price structure in such a way as would enable OPECs reference price of \$18/b to be achieved as soon as possible. It was decided that total OPEC production for the first half of 1989 should be set at a ceiling of 18.5 mb/d, distributed among individual Member Countries as follows:

	(million b/d)		(million b/d)
Algeria	0.695	SP Libyan AJ	1.037
Ecuador	0.230	Nigeria	1.355
Gabon	0.166	Qatar	0.312
Indonesia	1.240	Saudi Arabia	4.524
IR Iran	2.640	UAE	0.988
Iraq	2.640	Venezuela	1.636
Kuwait	1.037	*Total*	18.500

London, January 26, 1989 – OPEC/non-OPEC Technical Experts' Meeting: "In pursuance of the OPEC Ministerial decision in November 1988 to renew contacts with independent oil producers, Technical Experts of the Ministerial Group of OPEC met with their counterparts from independent oil producers. The OPEC representatives consisted of Algeria, Indonesia, Kuwait, Nigeria, Saudi Arabia and Venezuela.

The independent oil producers comprised the Province of Alberta in Canada Angola, China, Colombia, Egypt, Malaysia, Mexico, Norway, Oman, the Soviet Union, the States of Texas and Alaska in the US and the Yemen Arab Republic.

The Province of Alberta, the States of Texas and Alaska, the Soviet Union, Norway and Yemen Arab Republic attended as observers.

In a joint communiqué, the Experts took note of the positive market developments following the OPEC Agreement of November 1988, and exchanged views on the 1989 market outlook. The independent producers expressed their support for OPEC's aim of reaching a price of $18 per barrel for the reference basket as soon as possible, and both sides agreed on their joint responsibility to ensure market stability, for the benefit of both producers and consumers. They agreed on the need to take further steps to enhance the improvements achieved and to further strengthen the oil market, especially in the second quarter of 1989..."

85th, Vienna, June 5–7, 1989: "... Agreed that the reference price of the OPEC crude basket should be $18 per barrel, and that, accordingly, the OPEC production ceiling should be increased from 18.5 mb/d to 19.5 mb/d during the second part of the year and to distribute the increase among Member Countries on a pro rata basis as follows:

	(million b/d)		(million b/d)
Algeria	0.733	SP Libyan AJ	1.093
Ecuador	0.242	Nigeria	1.428
Gabon	0.175	Qatar	0.329
Indonesia	1.307	Saudi Arabia	4.769
IR Iran	2.783	UAE	1.014
Iraq	2.783	Venezuela	1.724
Kuwait	1.093		

86th, Vienna, November 25–28, 1989: "Decided to increase the OPEC production ceiling to 22 mb/d during the first half of 1990, and to adjust Member Countries' production levels accordingly:

	(million b/d)		(million b/d)
Algeria	0.827	SP Libyan AJ	1.233
Ecuador	0.273	Nigeria	1.611
Gabon	0.197	Qatar	0.371
Indonesia	1.374	Saudi Arabia	5.380
IR Iran	3.140	UAE	1.095
Iraq	3.140	Venezuela	1.945
Kuwait	1.500		

Vienna, March 16–17, 1990: "The 5th Meeting of the Ministerial Monitoring Committee decided to make no changes to existing production and pricing policies for the second quarter of 1990, and unanimously reaffirmed the 22 mb/d production ceiling and the $18 per barrel minimum reference price set in November 1989."

Geneva, May 2–3, 1990: "The Sixth Meeting of the OPEC Ministerial Monitoring Committee, specially convened to address the problem of the over-supply in the oil market and the consequent deterioration in crude oil prices, unanimously agreed to take immediate collective action to cut back OPEC production by 1.445 mb/d, the reduction to remain valid until the end of July 1990."

87th, Geneva, July 26–27, 1990: "After reviewing the oil market situation decided to make the minimum reference price for the OPEC crude basket $21 per barrel, and that the total ceiling for OPEC production for the second half of 1990 should be set at 22.4491 mb/d."

Vienna, August 8, 1990: "The OPEC Secretariat issued a statement in connection with the developments in the oil market following the invasion of Kuwait by Iraq on August 2. The President said fears expressed regarding possible disruptions in oil supply in the short term were the result of psychological and speculative factors rather than the manifestation of real market fundamentals..."

Vienna, August 29, 1990: "... Announced that in view of the uncertainty of oil supply in the fourth quarter of 1990 and the first quarter of 1991, the Organization had decided to increase production in accordance with need..."

Geneva, March 11–12, 1991: "The Ministerial Monitoring Committee, with a view to furthering the Organization's goal of achieving the OPEC basket reference minimum price of $21 /b, decided that OPEC's total production should be lowered to 22.3 mb/d for the second quarter of 1991..."

Geneva, September 24–25, 1991: "... Decided to raise OPEC's total production ceiling to 23.65 mb/d for the fourth quarter of 1991.

The Committee expressed OPEC's deep concern over the proposed energy tax being considered by the European Community. Oil was already disproportionately taxed at the consumption end: for the past three decades, consumers had been paying petroleum taxes which, in some cases, amounted to more than three times the delivered price of the crude oil..."

90th, Vienna, November 26–27, 1991: "Decided to retain the OPEC production level of 23.65 mb/d agreed upon in September 1991, and reaffirmed its commitment to the July 1990 Agreement, including the stated objective of $21 /b for the OPEC basket minimum reference price..."

Geneva, February 12–15, 1992: "The Sixth Meeting of the Ministerial Monitoring Committee decided that OPEC's production ceiling should, with immediate effect, not exceed 22.982 mb/d, including Iraq's current production of 505,000 b/d, and be distributed among Member Countries as follows:

	(million b/d)		(million b/d)
Algeria	0.760	Nigeria	1.751
Ecuador	0.273	Qatar	0.377
Gabon	0.273	Saudi Arabia	7.887
Indonesia	1.374	UAE	2.244
IR Iran	3.184	Venezuela	2.147
SR Libyan AJ	1.395	Kuwait	0.812

The Islamic Republic of Iran expressed reservations to the effect that a ceiling higher than 22.5 mb/d was a danger to the market. Saudi Arabia did not endorse its allocated production level of 7.887 mb/d and maintaned its offer to reduce by 500,000 b/d.

Geneva, September 16–17, 1992: "... In order to attain the minimum reference price of $21 per barrel for the Basket, OPEC's market share for the fourth quarter of 1992 should be 24.2 million barrels per day. The Committee reaffirmed that it would allow for any additional production from Kuwait and restated its commitment to the essence of the entire provision of the July 1990 Agreement."

92nd, Vienna, November 25–27, 1992: "Noted a downward trend in oil prices, with the OPEC Basket price falling from $18.80 per barrel in the first week of November. In order to arrest this trend and restore order in the market, the Conference unanimously agreed to a production ceiling of 24.582 mb/d for the first quarter of 1993, with the following allocations to Member Countries for the period January–March, 1993.

	(million b/d)		(million b/d)
Algeria	0.764	Nigeria	1.857
Gabon	0.293	Qatar	0.380
Indonesia	1.374	Saudi Arabia	8.395
IR Iran	3.490	UAE	2.260
Iraq	0.500	Venezuela	2.360
Kuwait	1.500*		
SP Libyan AJ	1.409	*Total*	24.582

* figure represents currents production level.

The Conference regretfully accepted the wish of Ecuador to suspend its full membership of OPEC."

Vienna, February 13–16, 1993: "... Noted the declining trend in oil prices which was due to an over-supply of oil in the market and a relatively low stock drawdown. The Committee decided to reduce the overal OPEC production ceiling to 23.582 mb/d, effective from March 1, 1993 allocated as follows:

	(million b/d)		(million b/d)
Algeria	0.732	Nigeria	1.780
Gabon	0.281	Qatar	0.364
Indonesia	1.317	Saudi Arabia	8.000
IR Iran	3.340	UAE	2.167
Iraq	0.400	Venezuela	2.257
Kuwait	1.600		
SP Libyan AJ	1.350	*Total*	23.582

Oman, April 13, 1993: "The Joint IPEC/OPEC Ministerial Meeting was attended by all 12 OPEC Members and 14 independent petroleum exporting countries, namely Angola, Bahrain, Brunei, Darussalam, the People's Republic of China, Colombia, Egypt, Kazakhstan, Malaysia, Mexico, Norway, Oman, the Russian Federation, the Republic of Yemen and the US state of Texas, some as observers. The Meeting reviewed the implication of the growing number of regional and energy blocs around the world, and noted increasing uncertainties in the long-term energy outlook. The Meeting also observed that petroleum products already carry a tax as high as $95 per barrel in some countries, and that further discriminatory tax increases in the consuming countries will have a destabilizing effect on the oil market. The Meeting called for increased co-operation between oil producers.

A Joint Working Group was set up to examine, *inter alia*, the long-term energy outlook and the impact of taxation on the oil producing countries."

94th (Extraordinary), Geneva, September 25–29, 1993: "... Expressed its concern both about the level of oil prices, which had deteriorated to such an extent that the OPEC Basket price had fallen below $15.0/b, and about the general instability in the market. In order to halt this slide and reverse the current downward trend to achieve a more reasonable price level, the Conference resolved to set the overall production ceiling for the following six months at 24.5 mb/d, distributed among Member Countries in accordance with the following allocations:

	(million b/d)		(million b/d)
Algeria	0.750	Nigeria	1.865
Gabon	0.287	Qatar	0.378
Indonesia	1.330	Saudi Arabia	8.000
IR Iran	3.600	UAE	2.161
Iraq	0.400	Venezuela	2.359
Kuwait	2.000		
SP Libyan AJ	1.390	*Total*	24.520

97th, Bali, November 22–25, 1994: "... decided to maintain until the end of 1995 the ceiling of 24.52 mb/d, distributed according to the temporary allocations decided in the 94th Meeting".

ANNEX 3

A MODEL TO EXPLAIN THE CONNECTION BETWEEN CRUDE OIL PRICES AND DAILY WORLD PRODUCTION OF CRUDE OIL

by Giorgio Spallanzani
EAI Hybrid Computer User Group, Noise Working Group, Concawe, Madouplein 1, B-1030 Brussels, Belgium

Introduction

The explanation of the relations between prices and supply of crude oil has always found an obstacle in the wide range of dollar prices for similar production value in different years. The following model attempts to introduce the time element, a variable z, which results in a spiral curve known as the "clothoid" or "the spiral of Cornu".

Description of the model

The model has the following equations:

$x = X(1)+X(2) \times Ch(z)$	in mb/d
$y = X(3)+X(4) \times Sh(z)$	in \$/b
$z = (u)^{X(5) \times X(6)}$	in non-dimensional time

these being the components of a parametric six-dimensional vector and u being the yearly variable with values of 0.15 for the first year (1973) with increments of 0.1 for each of the following years (non-dimensional).

The curve of Figure 8.1 (page 113) has many values on the y axis for the same value on the x axis.

This phenomenon can be interpreted in the following way. In the first phase, from 1970 to 1974, the oil prices grow together with production/demand, because consumers had been enjoying a "consumer's rent" and were ready to accept higher prices. From 1974 to 1979 prices grow much faster, and the production/demand curve becomes flat as the consumers look for alternative energy sources and practise energy conservation. Two different periods follow. The first, 1980–1982, has stable prices but a diminishing demand; the second, 1983–1993, has declining prices and a slow growing demand.

M. Colitti and C. Simeoni, Perspectives of Oil and Gas: The Road to Interdependence, 159–160.

The curve chosen for the model seems to describe this interaction between prices and production demand well, characterized by the rigidity of the latter and by a strong time-lag.

The model has not only more than one price for very similar values of production/demand in different years; it also has an asymptotic trend. It is in fact possible to think of time as a variable normal to the plane of the curve; increasing the time intervals, the differences between the value of y for each x tend to decline asymptotically for an infinite time, which, on the basis of the above-mentioned model, works out as follows

$$xlim = X(1)+X(2)/2$$
$$ylim = X(3)+X(4)/2$$

Result of the calculations

On the basis of the 24 historical data from 1970 up to 1993 a program for a non-linear optimization gave the following parameters:

$$X(1) = 48.00 \quad X(2) = 16.00$$
$$X(3) = 12.00 \quad X(4) = 21.00$$
$$X(5) = \ 1.54 \quad X(6) = \ 2.00$$

which correspond to the following limits:

$$xlim = 56.0 \text{ mb/d}$$
$$ylim = 22.5 \text{ \$/b}$$

The choice of values for u for each year has been made in such a way as to represent the rotation arch of the data.

Figure 8.2 (page 114) represents the empirical data available up to now.

M. Colitti and C. Simeoni, Perspectives of Oil and Gas: The Road to Interdependence, 159–160.

BIBLIOGRAPHY

9th World Energy Conference, Instanbul, 23–26 May 1977
10th World Petroleum Conference, Bucharest, 20–26 July 1979
12th World Energy Conference, New Delhi, 18–23 September 1983
12th World Petroleum Conference, Houston, 26–30 April 1987
14th World Energy Conference, Montreal, 17–22 September 1989
14th World Petroleum Conference, Stavanger, 12–15 March, 1994
15th World Energy Conference, Madrid, 20–25 September 1992
19th World Gas Conference, Milano, 20–23 June 1994

Abdelkader, Sid-Ahmed : L'OPEP – passé, présent et perspectives. Economica, Paris 1980
Adelman, Morris: The World Petroleum Market. Johns Hopkins University Press, Baltimore and London, 1972
Al Chalabi, Fadhil: OPEC and the International Oil Industry: a changing structure. Oxford University Press, 1980
Al Torki, Soraya and Cole, P. Donald: Arabian Oasis City. University of Texas Press, Austin, 1989
Al-Yousef, Ala'a: Kuwait and Saudi Arabia: from prosperity to retrenchment. Oxford Insitute for Energy Studies, 1980
Amuzegar, Jahngir: The oil story: facts, fiction and fair play. Foreign Affairs, July 1973
Annual Reports of Oil Companies, various years
API Basic Petroleum Data Book: petroleum industry statistics, various issues, Washington, DC.
APS Review, various years, Nicosia
Arab Oil & Gas Directory, various years, Paris
Arab Oil & Gas, various years, Paris
Askari, Hossein: Saudi Arabia Economy: oil and the search for economic development. Jai Press Inc., London, 1990
Atkins, James: The oil crisis: this time the wolf is here. Foreign Affairs, April 1973
Attiga Ali, Ahmed: The Arabs and the Oil Crisis 1973–1986. OAPEC, Kuwait, 1987
Attiga Ali, Ahmed: Interdependence on the Oil Bridge – Risks and Opportunities. Petroleum Information Committee of the Arab Gulf States, 1988
Beblawi, Hazem and Giacomo, Luciani: The Rentier State. Croom Helm, London, 1987
BP Statistical Review of World Energy, various issues. BP, London
CERA – Cambridge Energy Research Associates, Oil &Gas Watch, various years, Cambridge
CERA, Cambridge Energy Forum, various issues, Cambridge
Cedigaz, various years, Rueil Malmaison
Chadwick, Margaret, Long, David and Nissanke, Machiko: Soviet Oil Exports. Oxford Institute for Energy Studies, 1987
CIA, International Energy Statistical Review, various issues. US Government Printing Office, Washington, DC
Colitti, Marcello: Energia e Sviluppo in Italia. La vicenda di Enrico Mattei. De Donato Bari, 1979. Translated into Arabic by OAPEC
Colitti, Marcello: Energy and economic development. In: Development through Cooperation, 1981. Seminar between OAPEC and South European Countries, Rome, 7–9 April 1981
Colitti, Marcello: Size and distribution of known and undiscovered petroleum resources in the world, with an estimate of future exploration. In OPEC Review, Autumn 1981
Colitti, Marcello: Conditions constraints and tools for a cooperation policy between energy producer and consumer countries. 12th Congress of the World Energy Conference, New Delhi 18–23 Sept. 1983

M. Colitti and C. Simeoni, Perspectives of Oil and Gas: The Road to Interdependence, 161–166.

Colitti, Marcello: The oil industry view on cooperation between the developed and developing countries. Third Arab Conference on Energy and Arab Cooperation. Algiers, 4–9 May 1985

Colitti, Marcello: Managing the oil supply. OPEC Review, Autumn 1986

Davis, John: Libyan Politics. Tribe and Revolution. I.B.Tauris and Co., London, 1987

Davison, Ann; Hurst, Chris; Mabro, Robert: Natural Gas. Governments and Oil Companies in the Third World. Oxford Institute of Energy Studies, 1988

De Chazeau, M.D. and Khan, A.E. Integration and Competition in the Petroleum Industry. Yale University Press, 1959

De Golyer & MacNaughton: Twentieth Century Petroleum Statistics, various issues, Dallas, Texas

Deterding, Sir Henri: An International Oilman. As told to Stanley Naylor. Ivor Nicholson and Watson Limited, London, 1934

Economist Intelligence Unit, various issues, London

Emerson Steven: The American House of Saud. Franklin Watts, New York, 1985

Farouk-Sluglet, Marion and Sluglet, Peter: Iraq since 1958 – From Revolution to Dictatorship. I.B. Tauris and Co., London, 1990

Financial Times Energy Economist, various issues, London

First, Ruth: Libya – The Elusive Revolution. Penguin Books, London, 1974

Frankel, Paul H.: Mattei, Oil and Power Politics. London, 1966

Frankel, Paul (I. Skeet and R. Mabro, eds) Common Carrier and Common Sense – a Selection of his Writings 1946–1988. Oxford Institute for Energy Studies, 1989

Frankel, Paul: The Essentials of Petroleum. A Key to Oil Economics. Chapman and Hall, London, 1946 ; 2nd edition in 1969 by Frank Cass and Co., London

Fried, Edward and Trezise, Philip H.: Oil Security – Retrospect and Prospect. Brookings Institution, Washington, DC. 1993

Fried, Edward R. and Blandin, Nanette M. (eds): Oil and America's Security. Brookings Institution, Washington, DC, 1988

Hamilton, Adrian: Oil – The Price of Power. Michael Joseph/Rainbird, London, 1986

Hirst, David: Oil and public opinion in the Middle East. Faber and Faber, London, 1966

Holden, David and Johns, Richard: The House of Saud. Sidgwick and Jackson, London, 1981

Horsnell, P.: Oil Price Differentials: Markets in disarray .Oxford Institute of Energy Studies, 1990

Ibrahim, Ibrahim (ed): Arab Resources – The Transformation of a Society. Center for Contemporary Arab Studies, Croom Helm,1983

International Petroleum Encyclopedia, various years, Tulsa

Jaboli, Dante: Exploration for hydrocarbons in Italy. International Symposium in Tuscany, Florence, 1976

Jevons, Stanley: The Coal Question, ed. by A.W. Flux. August M. Kelley, New York, 1965

Julius De Anne and Mashayekhi, Afsaneh: The Economics of Natural Gas: pricing planning and policy. Oxford University Press, 1990

Lacey, Robert: The Kingdom. Hutchinson, London, 1981

Levy, Walter: World oil cooperation or international chaos. Foreign Affairs, July 1974

Lovins, Amory: Energy strategy: the road not taken? Foreign Affairs, October 1976

Luciani, Giacomo: L'OPEC nella economia internazionale. Einaudi, Torino, 1976

Luciani, Giacomo: The Oil Companies and the Arab World. Croom Helm, London, 1985

Luciani, Giacomo (ed): The Arab State. Routledge, London, 1990

Mabro, Robert et al: The Market for North Sea Crude Oil; Oxford University Press, 1986

Mabro, Robert (ed): OPEC and the World Oil Market – The genesis of the 1986 price crisis. Oxford University Press, 1986

Mabro, Robert: Political Dimensions of the Gulf Crisis. Oxford Institute for Energy Studes, 1990

Master, Charles et al: World Petroleum Assessment and Analysis, US Geological Survey, Reston, Virginia, 1993

Matar, Fuad: Sadam Hussein – The Man the Cause and the Future. Third World Center, London, 1981

Middle East Economic Surveys, various issues, Nicosia

Morse, Edward: The coming oil revolution. Foreign Affairs, Winter 1990–91

Mosley, Leonard: Power Play – The Tumultuous World of Middle East Oil 1890–1973. Weidenfeld and Nicolson, London, 1973

Nehering, Richard: Giant Oil Fields and World Oil Resources. Rand, St Monica, 1978

Niblock, Tim and Lawless, Richard: Prospects for the World Oil Industry. Croom Helm, 1985

Nore, Petter and Turner, Terisa (ed): Oil and Class Struggle. Zed Press, London, 1980

Odell, Paul and Rosing, K.: The Future of Oil. K. Page, New York, 1980

OPEC and Future Energy Markets – An OPEC Seminar, October 1979. Macmillan

OPEC Bulletin, various issues, Vienna

Parpinelli Tecnon, various years, Milano

Penrose, Edith: The Large International Firms in Developing Countries: The international petroleum industry. Allen and Unwin, 1968

Penrose, Edith: The Theory of the Growth of the Firm. Oxford University Press, 1966.

Petrocompanies Operating Data, various issues, London

Petroconsultants SA, various issues, Geneva

Petroguide Ltd, various years, London

Petroleum Economics Ltd, various issues, London

Petroleum Economist, various issues, London

Petroleum Information, various issues, Montagne-au-Perche

Petroleum Intelligence Weekly, various issues, New York

Petroleum Times, various issues, London

Pietra Italo : Mattei – La pecora nera. Sugarco, Milano, 1987

Platt's Oilgram News, various issues, McGraw-Hill, New York

Ratti, Giuseppe: Common Development and Cooperation between Countries of Southern Europe and the Arab World. In: Development Through Cooperation, Seminar Between OAPEC and South European Countries, Roma, 7–9 April 1981

Reviglio, Franco: The effects of the oil crisis on the Mediterranean region: a proposal to face increasing instability. Aspen Institute Italia, Istanbul, 22–24 June 1987

Robinson, Jeffrey: Yamani : The inside story. Simon and Schuster, 1988

Rustow, A. Dankwart and Mugno, F. John: OPEC Success and Prospects. Council of Foreign Relations, New York, 1976

Sarkis, Nicolas: Petrole e developpement economique dans le pays arabes. Presented to the Conference de le Centre d'Etudes Euro-Arabe, Dubai, 9–11 Janvier 1995

Sayigh, Yusif A.: Arab Oil Policies in the 1970s. Croom Helm, London, 1983

Sayigh, Yussif A.: The Arab Economy – past performance and future prospects Oxford University Press, 1982

Scott, Richard: Innovation in International Organization: The International Energy Agency. In International and Comparative Law Review. Inaugural Issue, Spring 1977, University of California Hastings College of Law

Seymour, Ian: OPEC Instrument of Change. Macmillan Press, 1980

Seymour, Ian: The World Refining System and the Oil Products Trade. Oxford Institute of Energy Studies, 1990

Shonfield, Andrew: Modern Capitalism. Oxford University Press, 1965

Simeoni, Claudio: Statistica Economica (for Eni's operators). Perseni, Rome, 1979

Simeoni, Claudio: Structural Changes in the International Oil Industry. Workshop of the Congressional Research Service, Library of Congress, Washington, 1984

Simeoni, Claudio: Hydrocarbon Resources in the Mediterranean Countries. Aspen Istitute Italia, "Mediterranean: World's Crossroads", Fondazione Cini, Venice, 1985

Simeoni, Claudio: Hydrocarbons in Europe and in Italy. ASEAN Seminar on Direct Applications of Natural Gas, Bandar Seri Begawan, Brunei, 1991

Simeoni, Claudio: Demand and Supply of Hydrocarbons in Europe to 2020. Unido, Vienna, 1994
Simeoni, Claudio: Impact on the European Gas Industry of Current Plans and Projects to Extend the Infrastructures. European Gas Strategies, Amsterdam, 1995
Skeet, Ian: OPEC – Twenty-five Years of Prices and Politics. Cambridge University Press, 1988
Spitz, Peter: Petrochemicals – The rise of an industry. John Wiley and Sons, New York, 1988
Stanislaw, Joseph and Daniel, Yergin: Oil: Reopening the Door: Foreign Affairs no. 4, 1993
Tétreault, Mary Ann: The Organization of Arab Petroleum Exporting Countries – History, Policies and Prospects . Greenwood Press, London, 1981
The Interdependence Model: Second volume of Development through Cooperation. Seminar between OAPEC and South European Countries, Rome, 7–9 April 1981
Treat, John Elting (ed): Energy Futures – Trading opportunities for the 1990s. PennWell Publishing Co., Tulsa, 1990
US Senate, SubCommittee on Multinational Corporations of the Committee on Foreign Relations Ninety-Third Congress: Multinational Corporations and United States Foreign Policy, 30 January 1974
Vernon, Raymond: Storm over the Multinationals: The Real Issue. Harvard University Press, 1977
Vernon, Raymond: Big Business and the State: Changing Relations in Western Europe. Harvard University Press, 1974
Vernon, Raymond: Technology Factors in International Trade. National Bureau of Economic Research, Inc., 1970
Weekly Petroleum Argus, various issues. Petroleum Argus, London
Wilson, Carrol: A Plan for Energy Independence. Foreign Affairs, July 1973
Worldwide Oil Directory, various issues, Pennwell, Houston
Yergin, Daniel: The Prize. Simon and Schuster, New York, 1991